The Right-Seat Handbook

The Right-Seat Handbook

A white-knuckle flier's guide to light planes

Charles F. Spence
Illustrations by author

TAB Books
Division of McGraw-Hill, Inc.
New York San Francisco Washington, D.C. Auckland Bogotá Caracas Lisbon London
Madrid Mexico City Milan Montreal New Delhi San Juan Singapore
Sydney Tokyo Toronto

Published by TAB Books, a division of McGraw-Hill, Inc.

pbk 4 5 6 7 8 9 FGR/FGR 9 9 8

Library of Congress Cataloging-in-Publication Data

Spence, Charles F.
The right seat handbook : a white knuckle flier's guide to light planes / by Charles F. Spence.
p. cm.
Includes index.
ISBN 0-07-060148-8 (pbk.)
1. Private flying. 2. Airplanes—Piloting. I. Title.
TL721.4.S63 1994
94.20183—dc20 94-20183
CIP

Acquisitions editor: Jeff Worsinger
Editorial team: Robert E. Ostrander, Executive Editor
Norval Kennedy, Book Editor
Production team: Katherine G. Brown, Director
Ollie Harmon, Coding
Brenda M. Plasterer, Desktop Operator
Jodi L. Tyler, Indexer
Design team: Jaclyn J. Boone, Designer
Brian Allison, Associate Designer
Cover design by John Martin, Mechanicsburg, Pa.
Cover photograph by Bender & Bender, Waldo, Oh.
Back cover copy written by Cathy Mentzer

GEN4
0601488

Dedication

To Majel
There are three women in every man's life:
the woman he loves,
the woman who loves him,
and the woman he marries.
I found all three in the same person.

Acknowledgments

Many persons inspired this book. They are the hundreds of thousands—spouses, family members, neighbors, and friends—who have so often asked of pilots: "Why do you spend so much time at the airport?"

Aside from these thousands, a thank you to Donald Koranda, who, although a dedicated flight instructor, resisted the temptation to urge more detail in this book than most non-pilots care to know while still making sure that in simplifying flight, the information remained technically correct.

Contents

Foreword

General aviation, defined as all aviation activity outside of scheduled air carriers and the military, has brought enormous benefit and pleasure to almost 700,000 licensed pilots who fly light airplanes. Scheduled airlines operate at fewer than 500 United States airports, while G.A. can land at almost 17,000 locations in this country. This opens horizons for travel and utility unheard of by any other means of transportation.

Weekend trips by car convert to a one-day affair through the use of a single-engine airplane. Complex whistle-stop business trips hitting several locations in a day become commonplace for the company using a corporate aircraft. In this age of time constraints, increased productivity, and lack of leisure time, the general aviation aircraft becomes a problem-solver.

Yet, flight in a light airplane can also create problems. Student pilots are unable to carry passengers, so they go through rigorous training, pass a written exam, and finish with both an oral and practical flight test. Only at this point, sometimes after more than a year of training, are they able to fly with passengers. The first to step cautiously into a "little plane" with a new pilot are often family or close friends.

I remember my first flight in Northern California with a freshly certified private pilot who wanted to demonstrate his newly acquired skills. It was a flight to lunch, some 50 miles from home. At the time I hadn't the slightest idea what made the airplane fly, or what all those gadgets on the dashboard had to do with staying in the air. All I understood from my right-seat vantage point was two gauges that were labeled FUEL; therefore, I would monitor them carefully. And I watched the pilot carefully for any physical or emotional signal that he wasn't sure of what he was doing.

My greatest fear was realized on the return trip when he must have been off course, and he started breaking out in a sweat while fumbling with a map that I had no way of understanding. It was shortly thereafter that I decided to earn my own wings, and throughout primary flight training it was evident that most formal flight instruction does not include any consideration or training for how to orient passengers, other than "Fasten your seat belt."

Years later, with almost a thousand hours of flight time, my constant right-seat companion and wife began expressing her feelings and offering me training in appreciating the nonpilot resource in the copilot seat. It took that long for her to speak up because up until that time today's modern intercom and headsets were not widely used. For almost 15 years I had flown without communicating, except for an occasional shout, with my passengers.

In the early 1980s, it became common practice to equip all seats with noise-reducing headsets with microphone attached. These suddenly opened up communication in a previously noise-laden environment. The advent of headsets also brought more demand on the pilot because now everything he or she said could be heard by others. For instance, my wife exclaimed that often I would voice an expletive as we were flying along, and she would immediately jump to the conclusion that our lives were over. Turns out all I had done was drop my pen and verbally reacted without consideration for the nonpilot in the right seat.

She has further allowed me to understand what it's like to sit through a long flight with absolutely no understanding of what's going on. Her reaction to this scenario has been negative because her mind would begin to conjure up all kinds of images and would wander from one flying disaster to another. Not so for the pilot, because he or she constantly has something to do.

Right-Seat Handbook by Charles Spence is a book that has been needed for decades—ever since airplanes had more than one seat. Unlike my wife, who finally decided to get her pilot's license just to be able to share in the experience, most people don't need to or have to go to that length. Throughout the following pages Mr. Spence demonstrates the tremendous support an "educated" right-seat passenger can provide.

The result is a positive cockpit environment and the active participation of two people, with perhaps only one a licensed pilot, to carry out the flight in a safe and efficient manner. Flying is challenging and rewarding, and *Right-Seat Handbook* makes it so for occupants of both front seats.

PHIL BOYER

Phil Boyer is president of the Aircraft Owners and Pilots Association.

Introduction

Snow, now dirty from days of gathering grime from vehicle travel, still covered the ground some six days after it had fallen. Chill winds sharpened the biting morning cold as I checked over the airplane and placed the two small suitcases in the luggage compartment. Satisfied of the safe condition of the craft, I climbed in and moved to the left seat. My wife followed, settling into the right seat. I reached across her, closed the cabin door, and made certain it was secure. That was the last of the winter winds we would feel for two days. The airplane was giving us the freedom to escape to a faraway land, if only for a weekend.

Snug in the single-engine Mooney, we climbed from the small airport just outside Washington, D.C. The first stop was Savannah, Georgia, for fuel and a quick lunch, then on to Daytona Beach. By midafternoon we had exchanged the wintry blasts of the Maryland snows for the sun and sand of Florida. We had a pleasant dinner at a seafood restaurant that evening, followed by more sun and surf the following morning. By Sunday night we were once again wrapping our bodies in heavy clothes against the Maryland winter, but rejuvenated for the coming week's work.

Sure, we could have gone by airline, but not on the spur of the moment. There would have been reservations to make and schedules to check, bound to departure and return when the carrier decided, not when we wanted. And the cost? Perhaps a toss-up. Two or even four people could go in the Mooney for the price of one person on a scheduled airline.

On another occasion, my wife, two sons, and I left the Washington area and in 45 minutes were at the airport near

Ocean City, Maryland. We enjoyed a full day on the boardwalk and the beach. Going and returning we crossed over the Bay Bridge that spans Chesapeake Bay and smugly commented about the traffic jams and the 3–4-hour journey those ground-bound travelers faced.

These are but two brief examples of what vistas the private airplane opens. At that time, neither my wife nor sons were pilots. (One son has since become a pilot with skills and talents much superior to mine.) This did not keep them from benefiting from the pleasures of travel by private airplane.

Besides such family trips, different group activities are available ranging from "let's all fly someplace for lunch today" to organized meetings of state pilot associations, flying club trips, "pancake breakfasts," or similar gatherings at various airports.

The Experimental Aircraft Association holds the biggest aviation gathering in the world every summer at Oshkosh, Wisconsin. More than 800,000 persons attend, and this event makes the Oshkosh airport the busiest in the world for the period.

Aircraft Owners and Pilots Association conducts an annual Expo convention. Rotated to various locations throughout the United States, this gathering offers not only aircraft exhibits and aviation programs, but evening entertainment, and special programs for the nonpilot.

Most of all, the private airplane extends the limits of activity. Distance has never been a concern. Time has been the limiting factor. Our great-grandparents thought nothing of taking three months to cross the continent. Now we get upset if we miss one section of a revolving door. The private airplane compresses time needed to cover greater distances.

But why the private airplane? Why not just take an airliner?

An airliner is great for 600–700 locations served by airlines—if the flights go when you want to go and return when you want. The private airplane puts you in command. It gives you access to almost 10 times as many airports that are open to the public and a near-equal number of private facilities, many of which may be used with prior permission. Many times these airports are closer to your ultimate destination. After all, very few times do you want to go to another airport; the airport usually is just an en route stop to your family's home, your business appointment, your day's outing location, or your vacation site.

Although there is a certain kind of person who loves airplanes just for the sake of flight, many others enjoy the results more than the flying. Usually the person who becomes a pilot does so because of the pure enjoyment of flight and, to some extent, the satisfaction of achieving a distinction that not everyone earns. Whether or not they admit it, many pilots find the travel and the extension of horizons for living and work as bonuses to their initial passion for flying.

Passengers can skip the learning process and go directly to enjoying what the airplane offers.

It is not necessary to be a musician to enjoy good music. But it helps to have a smattering of knowledge. So, too, with flight. Knowing something of how the airplane works, what the pilot is doing, and what you, as a right-seat passenger, can do to help, will enhance your pleasure of getting to the places the airplane can take you. Also, if that pilot in the left seat happens to be your spouse, parent, other relative, friend, or business associate, your knowledge and understanding of flight will not only improve the atmosphere of the trip, but will help alleviate short tempers and hurt feelings by avoiding unnecessary questions followed all too often by curt answers.

The information in this book will not make you a pilot. It might make you *want* to become one, but that is a different story. What you will be after reading this book is a better-informed passenger who has lost most fears of flight and has learned to be an extra hand in the cockpit.

Your favorite pilot would probably say that the information in this book is too simplistic: "There's much more to good, safe flying than what appears here." That's true. For anyone who wants to become a pilot there are ample books, lectures, courses, schools, and instructors to impart the knowledge, the training, the testing to achieve that goal. That's not needed to gain more enjoyment as a passenger.

If the pilot with whom you fly is the macho superhuman who likes to boastfully proclaim that piloting is only for the anointed few, please put a plain cover on this book as you read it. Seeing that the favorite passenger is discovering the simplicity of flight, the ease of understanding navigation, and the safety of the well-maintained airplane might burst the bubble of pride that surrounds some fliers.

1

Getting there can be half the fun

And do you promise to love, honor, and enjoy flying in a light airplane?

Besides the bank that holds the mortgage and perhaps a family member questioning why the pilot spends so much time at the airport, four forces act on an airplane: lift, thrust, drag, and gravity (weight) (Fig. 1-1). Nonpilot passengers in the right seat thoroughly understand the power of gravity, but some find that a lack of knowledge about the other three often leads to unnecessary concerns that distract from the enjoyment of getting to the destination.

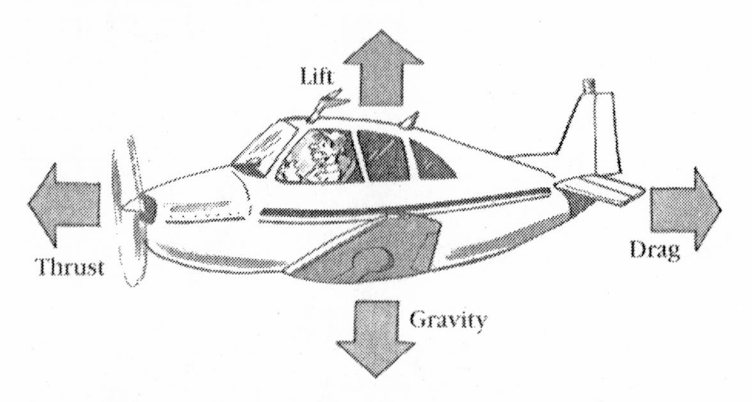

1-1 *Flying an airplane is the ability to properly and safely control the four forces that affect flight.*

Gravity is that force that has kept the human an earth-bound, two-dimensional creature since the beginning of time. Another force, *lift*, is an equally valid principle of physics, and it did not take an apple falling on the head of a scientist to understand it. An eighteenth century scientist named Bernoulli discovered what has become known as—imagine this!—Bernoulli's Principle.

How the airplane flies

To know what this has to do with enjoying flying, we must look at the environment in which we live. On Earth, we are at the bottom of an ocean, an ocean of air. We breathe it. We see results of its movement in swaying trees. Sometimes we might even think we see air when the smog settles in. We know it's there but rarely think about it. In aviation, this air ocean must be considered because it is the environment in which flight takes place.

Bernoulli discovered that the pressure of a fluid, either liquid or gas, decreases at a point where the speed of the liquid increases. Hold that thought. Now look at the shape of an airplane's wing (Fig. 1-2). It is bigger in the front than in the back. (Pilots call these the *leading edge* and the *trailing edge.*) As the wing moves through the air, that big front edge disturbs the airflow. To travel up and over the leading edge of the wing, air must move faster; thus, Bernoulli tells us that this reduces pressure. There is stronger pressure below, where the air is not moving as swiftly. Some purists argue that the wing moves in to fill the area of decreased pressure; others contend that a stronger pressure below pushes the wing upward. Who cares? Either way, the scientific principle of lift is just as valid as is the principle of gravity.

1-2 *The shape of the wing causes air pressure to change, which creates lift.*

Prove this to yourself. Take a strip of paper 4 or 5 inches long and about 1 inch wide. Tissue paper works best, unless you can blow very hard. Hold the paper to your bottom lip, and blow over the *top* of it. Blow hard, and watch the paper try to rise (Fig. 1-3). The harder you blow, the faster the air travels over the paper; the softer you blow, the less difference

1-3 *Blowing over the top of a piece of paper reduces air pressure, causing the paper to rise.*

in pressure, and the paper begins to fall. This is why pilots are concerned about airspeeds. Enough airspeed is needed to assure that the force of lift is greater than the force of gravity.

In addition to speed, the angle at which the wing passes through the air also determines the amount of lift that is created. If a wing is placed at too steep an angle, the air that is flowing over the top can no longer follow the wing's curvature. Instead of creating lift, the air begins to "burble" behind the wing, and lift is destroyed (Fig. 1-4). This is called a *stall.* This has nothing to do with a failure of the engine, as in an automobile. When a wing stalls, it loses lift, and gravity becomes the controlling force.

A third controlling force is *thrust.* Thrust is the power that propels the aircraft. On propeller-driven aircraft, thrust is generated by turning blades biting through the air much as the threads on a screw move through fibers of wood, or as a corkscrew easily moves into a cork. A jet engine creates thrust with a powerful backward push on the air, or by creating power that turns gears connected to a propeller shaft. As a

1-4 *A wing that is placed at too steep an angle prevents air from flowing over the surface, which causes the wing to "stall," losing its ability to create lift.*

right-seat passenger, it makes no difference which kind of thrust powers your aircraft, except to determine the speed of your trip.

Drag is the fourth force acting on the airplane. The airplane moving through the ocean of air creates friction. Pilots have little control over friction, except to keep the airplane clean and determine what kind of antennae or other devices hang on the airplane.

What if the engine quits?

The pilot controls the aircraft by controlling the four forces of gravity, lift, thrust, and drag. In the most simplistic terms, takeoff and climb are accomplished by inducing power and manipulating the controls so that thrust and lift overcome gravity and drag. The pilot achieves descent and landing by permitting gravity and thrust to become the more dominant forces.

Notice that in both cases, thrust is important, which leads to the question: "What happens when the engine stops?" Usually that's good because it means the flight is over, the airplane is safely parked, and the pilot and passengers can finally go to the bathroom. On the extremely rare occasions when the engine stops while flying, there is still strong forward thrust that can keep the airplane moving while the pilot looks for an open space to land. This thrust, however, comes from the pull of gravity, which can be controlled.

If the engine quits in a typical single-engine airplane, the pilot can adjust the controls so that the aircraft, without power, will descend at a rate no greater than about 500 feet a minute. Let's say you are flying at 5,500 feet above the ground. That means at a descent rate of 500 feet per minute, the pilot has 11 minutes to search out a good, safe place to land. There is a gliding ratio that reveals how many feet forward the airplane will travel for a given number of feet of descent. Most pilots don't even know this formula, so there is no need for nonpilots to worry about it.

Are two better than one?

If the chances of one engine failing are so remote, and an airplane turns into a glider if it does quit, why do some persons spend the extra money to have two engines? There are many reasons.

Lifting power is an important benefit. A 250-horsepower engine can carry a given amount of aircraft and its load of persons and baggage. Put on a second 250-horsepower engine and it can carry more. The aircraft also can be larger and roomier for more comfort. The twin-engine airplane usually gives the owner greater range, more speed, and more options for juggling load and range.

A person who flies frequently when visibility outside the aircraft is limited, doesn't want the airplane to become a glider with less maneuverability if one engine has problems. Persons who fly often at night also appreciate the extra margin of safety offered by a second engine. Whatever the conditions, it is comforting to know that if the remote possibility occurs and there is a problem with one engine, another one is out there purring away to maintain altitude and fly to the destination or a safe landing at a nearby airport.

I recall well a flight with my wife and two sons in a single-engine Piper 235 Pathfinder in solid instrument conditions. We were en route from Maryland to Cincinnati, Ohio. Weather at either end of the trip was clear with only light winds; however, solid clouds were over the Allegheny Mountains as usual. I filed an instrument flight plan, and I was assigned an 8,000-foot cruising altitude.

As we milled along in the clouds like inside a milk bottle, the engine hummed a perfect rhythm. Below were the tree-covered mountains, which I could not see. Suddenly, the engine coughed. Perhaps a piece of dirt moved through the fuel line. My thoughts immediately jumped to wishing I had a second engine if this were to be more than a quick hiccup. That one spit of the engine kept me focused on our exact position, the elevation of the mountains below, and the distances to the nearest airports until we came out of the milk-bottle weather over eastern Kentucky and I saw the ground flattened below us.

A second engine is not an absolute assurance of reliability, however. Most light-twin airplanes cannot fly as high on one engine as they do with both operating. It's simple logic. Take away half of the power and the aircraft cannot keep its same lifting ability. The density of air decreases as altitude increases; thus, every airplane has a *service ceiling*, which is the highest altitude that it can fly. A service ceiling for a twin-engine airplane is that altitude at which it can get a 50-foot-per-minute climb with the critical engine not operating and the aircraft at gross weight. Ability to climb is necessary if the air is turbulent. The *absolute ceiling* is the highest attainable altitude at gross weight.

Lets's say you are flying along as the right-seat companion and the pilot has the aircraft at 10,000 feet. You are above the weather, in a comfortable light-twin airplane that has a single-engine absolute ceiling of 7,100 feet. If one engine is shut down for any reason, the aircraft will descend from the 10,000-foot cruising altitude to no higher than 7,100 feet. Even when flying over the Rocky Mountains where the ground level is higher than the service ceiling of many light-twin aircraft, the second engine provides power to maneuver to a safer site for a forced landing.

These comments about the limitations of light-twins are intended to confirm the reliability and safety of single-engine aircraft and suggest that two is not always better than one. What the two do offer, however, is an opportunity to gain even more of the benefits that travel in your personal airplane offers.

Just as the models of single-engine airplanes differ while offering a variety of choices for the pilot, so, too, do multiengine aircraft present a choice of vehicles to match the needs and financial resources of pilots and owners.

Some twins, such as the Beechcraft Baron and the Piper Seneca, have cabin designs similar to those of single-engine aircraft. Pilots and passengers enter through a single door, and usually the seats are arranged with limited walking space between them.

Cabin-class twin-engine airplanes have a larger interior typically with walking space between two rows of seats. The door usually has stairs for entry and exit. Amenities such as curtains close off the flight deck from the main cabin, and often a chemical toilet can be curtained off. More space and larger engines mean more baggage may be carried onboard.

The larger twin-engine airplanes can have more of the comfort that nonpilots are familiar with from traveling on airliners, perhaps more luxurious. These include such niceties as sofa-seats, refrigerators, bars, fold-away tables, communication ability between the flight deck and the cabin, stereo systems, air conditioning, and pressurization.

Some of the high-performance, single-engine aircraft also offer one or more of these comfort-and-convenience items, but this is the exception in the smaller end of the product line.

A few single-engine aircraft, like the Piper Malibu or a certain model of the Cessna Centurion, are *pressurized*, but this convenience is more likely found in a cabin-class twin. Pressurization provides more comfort at higher altitudes, permitting the aircraft to cruise at altitudes that are often above adverse weather and where faster speeds are usually attainable. The pressurization system keeps the occupants at a warmer and more comfortable air pressure found at a lower altitude.

Most light twins are powered by *piston engines* similar to those used in the single-engine class. Many of the engines are essentially identical. When moving into the cabin-class, however, the purchaser has a choice of models that offer either piston or *turbine engines.* (The piston engine operates like a car engine and burns a fuel similar to gasoline. A turbine engine burns a kerosene fuel, and the internal parts are completely different than a piston engine.) The ultimate in personal flying, of course, is the corporate jet. Cost of purchase and operation keeps these birds out of reach for the average traveler. The jet has proven itself as a practical and profitable business tool that gives companies a travel edge over competitors that rely on the limited scheduling and service points of commercial airlines.

Whatever kind of airplane is used for travel, the principles of flight are the same even though techniques and pilot-skill requirements might differ in minor ways. With this in mind, let's see what makes an airplane fly and how the pilot, with a little help from the right-seat companion, goes about the enjoyable task of air travel.

Controlling the four forces

The same four forces apply to everything that flies, from a paper airplane to the space shuttle. Whether a single-engine or a twin-engine general aviation airplane (Fig. 1-5), or a massive airliner, the aircraft reacts to how the four forces are controlled by the pilot.

Thrust is easy. That comes from the engine and the propeller. A throttle controls the engine as does an accelerator on a car. On an airplane the throttle is controlled by a hand lever instead of a foot pedal. Push it in for more power; pull it back for less power. On many airplanes, the angle of a propeller's blades is controllable, permitting increased bites in the air for takeoff and climb. (To the passenger in the right seat, this knowledge is important only to avoid the smug looks a pilot gives when asked why there are two knobs to push and pull.)

Thrust alone, however, is not enough to make an object fly. Unless thrust is combined with something to control the

1-5 *Typical general aviation airplanes include (top to bottom): a French-made Tobago with a low wing; a Cessna Skyhawk with a high wing; a Piper Aztec light twin-engine airplane; and a Beechcraft King Air cabin-class twin.*

flow of air over the surfaces and create lift, the vehicle becomes nothing more than a funny looking car that threatens other aircraft and hangars on the airport.

When we look at the tail section of the airplane, one part is horizontal to the ground (Fig. 1-6). On some airplanes, just the rear portion moves; on others, the entire section moves. When only a portion moves that portion is called *elevators*; when the entire portion moves, it is called a *stabilator*. The *control wheel*, or *stick*, controls this surface of the airplane. Pull back on the wheel, and the elevator moves upward (Fig. 1-7); hence, the moving air hits it and pushes the tail section downward. If the tail of the airplane goes down, the front must go up. This, coupled with lift generated by air passing over the wings, lifts the airplane off the ground.

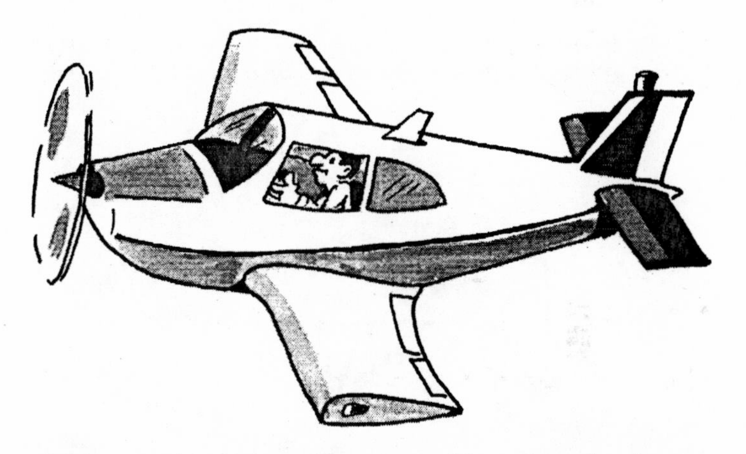

1-6 *The movable, horizontal sections of the tail are the elevators.*

To descend, the position of the elevator is reversed by pushing forward on the control wheel or stick. This reverses the position of the elevator so that air strikes it on the bottom, causing the tail to rise and pushing the nose downward (Fig. 1-8). Of course, management of power is involved in both climb and descent, but this is a concern of the pilot, not the right-seat passenger.

While we are at the tail section of the airplane, look at the portion that is perpendicular to the ground. This portion is the

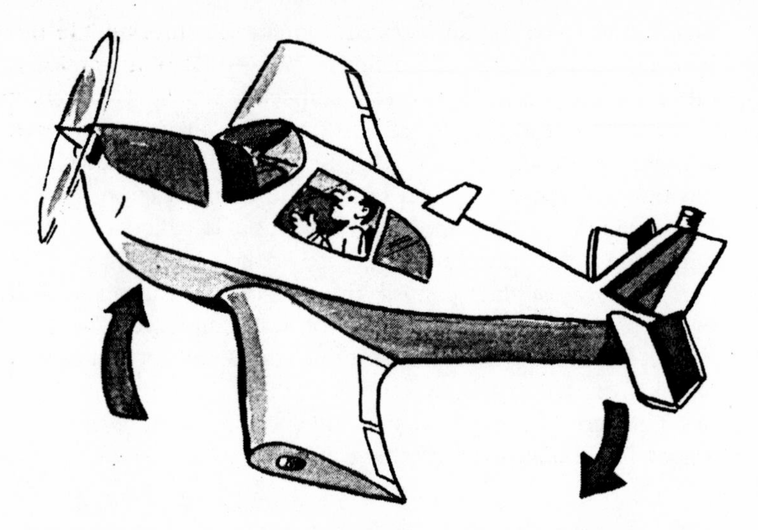

1-7 *When the elevators move above a level plane, wind strikes them, forcing the tail section down, which causes the nose to rise.*

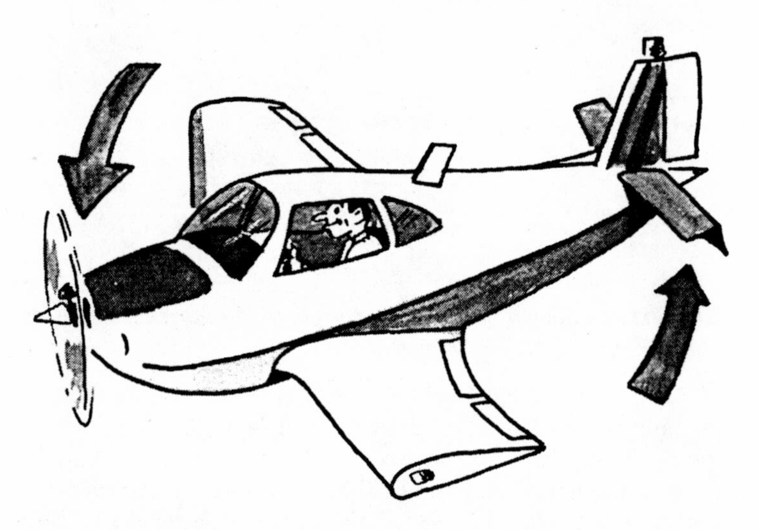

1-8 *When the elevators are placed below the level plane, wind strikes the underside, forcing the tail section up, which causes the nose to drop.*

vertical stabilizer and *rudder*. The part that moves is the rudder. Pedals located at the feet of the pilot control the action of the rudder. The rudder operates similarly to the elevators in that when moved, air strikes one side more than the other, pushing the airplane's tail in one direction and causing the nose to move in the opposite direction (Fig. 1-9).

1-9 *The rudder, the movable section of the vertical part of the empennage, helps to smooth a turn.*

Contrary to popular belief of nonpilots, the rudder is not the primary control for turning the airplane. An airplane will turn, or change its heading, whenever the wings are moved from a level position into what is commonly referred to as a *bank*. To place the aircraft into a *coordinated turn*, the pilot must use two controls, the ailerons and the rudder. The rudder is used in a supporting role to counteract some of the aerodynamic forces working against the turn and to make the turn smooth and comfortable for everyone in the aircraft.

Look at the wings. The *ailerons* are the movable parts at the back of each wing, away from the fuselage, toward the wingtips. By turning the wheel or moving the stick, the pilot causes one aileron to rise above the top surface of the wing, and the other aileron on the opposite wing moves lower than the bottom of the wing (Fig. 1-10). Remembering what we learned about lift being created by the speed of air over the wing, changing this shape causes one wing (the wing with

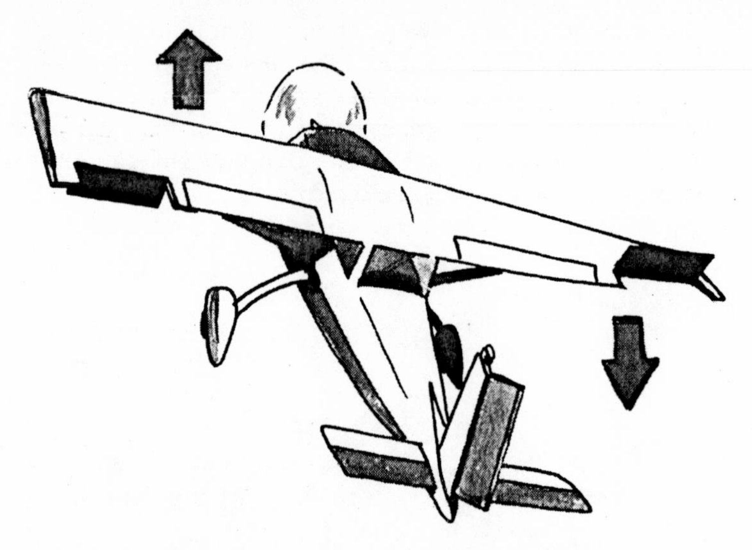

1-10 *Ailerons, located on the outer sections of the wings, change the air pressure by moving above or below the even plane of the wing causing one wing to have increased lift while the other has decreased lift. This causes the airplane to "bank," or turn.*

the lowered aileron) to have more lift that the other wing that has the heightened aileron. This puts the aircraft into a turn, or bank.

With just ailerons to make a turn, the airplane would *slip* around. By coordinating the rudder and ailerons, the pilot makes a smooth turn. Some back pressure must be held on the wheel or stick to compensate for decreased lift on the wings when they are in a turn.

Looking at the trailing edge of the wings, you will see larger, movable sections that are located closer to the fuselage. These are *flaps* (Fig. 1-11). Lowering flaps provides increased lift to permit safe operation at slower speeds. Some classic airplanes such as the Piper J-3 Cub or Aeronca Champion do not have flaps. Flaps are used primarily during the approach for landing, but some aircraft also use them for takeoff.

Although the theory of flight seems easy to understand, it took the human millions of years to conquer it. The Wright brothers finally achieved *controllable* flight. They and others

1-11 *Flaps are extended below the even plane of the wing, which increases the speed of air moving over the top of the wing, resulting in increased lift and permitting safe flight at slower speeds.*

had built gliders, but Orville and Wilbur were the ones who proved the idea to control climbs, descents, and turns by changing the flow of air over the surfaces of the aircraft.

Qualifying the pilot

This simplistic explanation of flight should help allay any fears that the right-seat passenger might have about the validity of flight and the safety of the airplane. If this isn't enough, everything about aviation is approved by the federal government. To some, it might not be a source of security to know that "I'm from the government, and I'm here to help you," but in aviation, the Federal Aviation Administration plays a major role in keeping flight a safe form of enjoyment and transportation. From the initial design, through every rivet and instrument on the aircraft, certain federal standards must be met. This is true not just for the original manufacturing of the aircraft and installation of equipment, but for maintenance, also. The aircraft must be thoroughly inspected at regular intervals; most work may be conducted only by federally certificated technicians. Even the airports from which the airplanes operate are inspected and meet certain federal or state requirements.

All this safety consideration extends to the pilot. Unlike boaters and auto drivers, pilots must obtain a specified amount of training. They also must practice alone over many hours before taking passengers. The FAA conducts extensive testing through written examinations that cover such subjects as theory of flight, weather, Federal Aviation Regulations, navigation, communication, and proper loading of the aircraft for weight and balance.

When this extensive written test is passed, the pilot must also pass an oral examination and a flight evaluation given by a federally designated examiner. But that's not the end of it. After a pilot obtains a certificate, certain recency of experience levels must be maintained within a three-month period to carry passengers or to operate the aircraft at night. If rated for flight by reference to instruments only, additional recent experience is required. Every two years the pilot must take a flight review with a certificated instructor and again be questioned on operations, rules, and other subjects of flight and demonstrate proficiency at manipulating the controls for safe flight.

As if this isn't enough, the pilot must see a doctor at least every two years for a physical examination to test vision, blood pressure, hearing, and other bodily functions that affect flying safety. Every visit to a physician between examinations must be reported, even if the purpose was something as nonflight impairing as removal of a hangnail.

A major key to safety is making certain that the flight does not exceed the capabilities of the pilot and the airplane. Smart pilots know that when weather is bad or some other condition makes the flight potentially dangerous, it is better to be on the ground wishing you were flying than to be flying wishing you are on the ground.

With this reassurance of safety, go ahead. Get into that right seat. You will find that getting there *can* be half the fun.

2

Look before you leap

Don't bother your mother. She said that she needs to study the clouds before she flies us to your uncle's house.

Getting into an airplane, even for a short flight, is much different from hopping into the car or even saddling up the old riding horse. There is more preparation than sticking a car key into an ignition switch and pulling out of the driveway; however, considering that the small airplane takes you cross-country two, three, or four times faster than the automobile can do it, a few minutes spent in advance planning is of little significance. It's also one of the reasons why travel by private airplane is safe.

Pilots begin flight planning, or *preflight preparation*, hours or even days before starting on the trip. The first step in preflight planning is to determine the general routes available, examine the terrain, obstacles, and obstructions along the route, and watch the developing weather patterns. Looking in more detail at specific aspects of the weather will help to determine which routes to take and what to expect along the way.

There's a saying about pilots. You can always identify one. The pilot has a big wristwatch on his arm, a pair of sunglasses attached to his belt, and a bus schedule in his pocket. That might be only partially true; some pilots carry their sunglasses in pocket cases. One can, however, identify a pilot by the way he or she looks at the sky. Weather is a major factor in flying, and good pilots study it conscientiously.

Weather can make the flight anything from impossible to thoroughly enjoyable. Wind can make a flight longer or shorter depending on which way it is blowing. Like going with the current of a stream, going with the wind moves the airplane at the speed the wind is blowing. Going the other way, wind slows a flight.

Because the airplane is riding in a sea of air, if that air is moving at 30 miles an hour, and the airplane is traveling the same direction that the wind is blowing, the airplane is passing over the ground 30 miles per hour faster than its speed through the air. An airplane moving through the air at 130 miles an hour suddenly is moving over the ground at 160 mph.

Go the other way, and the opposite is true. Against the wind, that 130-mph airplane travels over the ground at only 100 miles per hour. Pilots call a wind blowing the airplane along a *tailwind*. Now, guess what one that is blowing on the nose of the airplane is called? No, not a "nosewind." It's a *headwind.* How logical!

Just as fathers tell their children that when they were young they had to walk through 2 feet of snow going to school—and uphill both ways—some pilots claim they have a built-in headwind, but don't believe it. Wind changes directions. If a trip takes you through a weather front, wind direction will change during this one flight.

It doesn't take a rocket scientist to figure out that wind can make a big difference in the amount of time the trip takes, and thus the amount of fuel that will be used. It can affect whether an en route stop should be made or if arrival might be after dark, or after weather at the destination is forecast to change. It can also mean that, with a good tailwind, you'll arrive at your destination hours before Aunt Minnie and the cousins are at the airport to meet you.

Checking the weather

There is more to preflight planning than winds. Clouds give the pilot a message. They can signal advancing weather patterns. They can tell the pilot whether the flight will be smooth or bumpy. They can foretell the approach of a weather front, which is that line between a mass of warm air and a mass of colder air. A good, safe pilot will know enough about clouds to fly safely in them, if the pilot is trained and licensed for instrument flight and the aircraft is properly equipped. Or a good pilot will know if the flight can be made below the cloud layer, or by taking paths around clouds. Other information about clouds might make a safe pilot decide not to fly at all.

As a passenger in the right seat, you might not care to know everything a pilot must know, but here are a few bits of information that will make the pilot realize you are more than a casual weather observer.

If the clouds are billowy and rising in tower-shapes (*cumulus*), it's a good indication that the air is unstable because of rising air currents. Under these clouds, the air probably will be turbulent so your trip will be "bumpy." The airplane will bounce more. Cumulus clouds usually form below 6,000 feet, so the pilot often will be able to climb above these clouds where the air is smoother, and the ride is more pleasant. Cumulonimbus, on the other hand, can build up into severe thunderstorms and reach heights of 40,000–50,000 feet or higher in severe storms.

Stratus clouds form in layers. These indicate a more stable air mass, so your ride will be smoother in the air below them. Visibility usually is good below the cumulus-type clouds but is poorer beneath stratiform clouds where they form nearer the surface causing low ceilings.

In the Northern Hemisphere, weather moves generally in a west-to-east pattern, so the old sea adage of "red sky at morning, a sailor's warning, red sky at night, a sailor's delight" applies to flying as well. A red sky at evening would generally indicate that to the west, where the sun is setting, there are no clouds, so the sunset is visible.

Pilots also want to know if there might be precipitation that could cause reduced visibility, icing, hail, or snow. The pilot wants knowledge of weather not only along the route, but also beyond the route to determine which way and how fast the weather is moving.

Pilots aren't meteorologists (although some like to think they are), but many meteorologists are pilots. A good and safe pilot will know enough about weather to fly safely in it, or make the proper decision not to fly at that time.

The pilot must know basic weather information to receive an FAA pilot certificate. The *Federal Aviation Regulations* require that the pilot obtain all pertinent weather information before beginning a flight.

If you are a family member of a pilot, this obsession with weather before a flight might seem unnecessary, but at 5,500 feet it is difficult to pull to the side and let a storm pass.

The pilot has several sources for finding weather information. For general patterns, the nightly news, one of the morning television shows, or "The Weather Channel" on cable offer a good overview. Thanks to support from different organizations, many PBS television stations carry a special aviation weather program each weekday morning. "A.M. Weather," produced by Maryland Public Broadcasting, gives a detailed report (Fig. 2-1). Some pilots have found it helpful to use a Polaroid camera and snap pictures of pertinent graphics on these tele-

2-1 *Meteorologists Joan Von Ahn, Carl Weiss, and Wayne Winston are the "A.M. Weather" broadcast team that offers weather information featuring special aviation reports every weekday morning on PBS stations nationwide.*

vision programs. This supplies a record to take on the flight, and gives a better understanding of what to expect and what to ask for when the complete weather briefing comes.

This still is not enough for the pilot. Those with computers can obtain full weather briefings from different sources, some commercial, charging for each report, others free, supported by the FAA. The DUAT (Direct User Access Terminal) service is funded by the FAA for basic weather information. Providers of the service, such as GTE Federal Systems, also offer additional flight information for pilots (Fig 2-2). Information gained through the computer can be saved as hard copy from a printer that the pilot may take along on the flight.

GTE Government Systems Corp.

***2-2** The DUAT service is available by computer either at home base or with portable laptops to use at en route stops. GTE Government Systems is one company offering DUAT under contract to the FAA.*

The Federal Aviation Administration maintains about 175 flight service stations throughout the country. The flight service station has many functions, but providing weather briefings is what your pilot is concerned about first. These briefings are by telephone or personal visit if the location of the FSS and the pilot are compatible.

Planning the route

When the pilot has determined the weather, he or she can decide the route of flight. To go from Chicago to New Orleans, for instance, a nonpilot might say "just fly south until you see a big body of water with a large city near it." That's great if the weather is fine all the way, but if there are thunderstorms over St. Louis, for example, the prudent pilot might plan a flight west to Des Moines, down over Kansas City, and then begin to angle southeast behind the bad weather to New Orleans. This might be 200 or so miles farther, but in an airplane that does 145 miles per hour, instead of lasting about 5 hours, the trip requires another hour and 22 minutes. Not a bad trade-off to get a smoother ride with greater safety.

After getting the best weather information available, the pilot begins to spread out all those charts and papers on the kitchen table. This is more than taking a road atlas and sketching out highway numbers. The pilot charts the course on either a *sectional chart*, used primarily for flight under visual flight rules (VFR), or an *en route low-altitude* chart, used primarily when flying under instrument flight rules (IFR). (Don't worry about these differences now; we will discuss finding your way later.)

The flight plan will break the trip into various segments including distances and estimated flight time between points. (These segments are called *legs*, which is rather silly because we are using wings and not legs. Perhaps each segment should be called a "wing." But we must remember that this language is the same one that calls material sent by ship "cargo" while that put on a truck is a "shipment.")

Breaking the total trip into short wings—excuse me, legs—provides essential information to judge fuel consumption and general progress of the flight.

The pilot also will calculate *weight and balance*. This takes into account the weight of the persons, the baggage, and the fuel to be carried, and where to place all this in the airplane. After all, as remarkable as the power of lift might be, the size of the airplane and engine determines just how much can be taken aloft. If total weight exceeds the limits of the aircraft, the pilot might either decide to reduce the amount of fuel (which weighs about 6 pounds per gallon), or leave something or someone on the ground. A reduced fuel load just means that additional stops might have to be made along the way.

All of this usually occurs before going to the airport, or well before loading the luggage and kids into the airplane.

Conducting the airplane preflight

Once the flight planning is completed, it's now time to make sure the airplane is in top condition. This is done by conducting a *preflight inspection* of the airplane.

You'll see the pilot go out to the airplane and start a close examination of all parts of the airplane that can been checked without dismantling the airplane. Ask what the pilot is looking for, and the answer will probably be "Anything that doesn't look right." That's not a smart-aleck answer to put down the nonpilot. In a preflight inspection, the inspector really doesn't know what he or she is looking for. It can be anything that isn't right—and that won't be known until it is seen.

The pilot usually has a set routine for doing the preflight. First, it is to check the interior of the airplane to make certain that all switches are off and the parking brake is set. From here the inspection goes down one side of the airplane, checking for any damage. It's possible that another aircraft has clipped a wing, perhaps the lawn mowing crew plowed into the tail section, or maybe the gas-hose nozzle was dropped on the wing.

Around the back of the airplane, the pilot will look at the hinges, nuts and bolts, and general condition of the rudder, the elevator, and the stabilizer. If it is an airplane with the tailwheel, this will be checked. Then around the other side up to the wing where the flap and aileron are inspected. The main landing gear are inspected for damaged or worn tires, properly connected brake lines, and the general condition of the wheels.

Pilots are skeptical persons. Although the fuel gauges might read that the tanks are full, all pilots are from Missouri and insist upon "being shown"—respecting the state motto. They will remove the caps and visually check the quantity of fuel.

The look into the tanks might confirm that they are full, but the pilot wants to know more: Full of what? Different engines require different grades of fuel. These grades are dyed different colors. Also, air inside the tanks might condense as temperatures change in much the same way as it does on the outside of an iced-tea glass, forming water droplets that collect in the fuel.

That's why the pilot opens a drain and takes a sample of the fuel. It must be the right color, show no visible water, and be free of contamination. Several samples might be taken until the pilot is satisfied that the fuel is safe.

Moving around to the engine, the pilot opens the *cowling* to look inside. The oil level is checked, and the pilot makes sure that all hoses and wires are in place. There is another reason for opening the cowling. Birds and small varmints of the field find the covered engine area an excellent place to build nests (Fig. 2-3). A determined bird can build enough of a nest within a few hours to present a serious fire hazard.

2-3 *A preflight inspection under the cowling sometimes can turn up unexpected results.*

The pilot goes around the other wing, checking another fuel tank (usually there is one fuel tank in each wing), and checking the other aileron and flap. The preflight inspection is complete when the pilot is back at the cabin door. Any discrepancy can cause the flight to be canceled or delayed until the problem is corrected.

"Hey," you say, "what about the engine? Just because it looks right and the bluebird of happiness hasn't decided to set up housekeeping, how do we know it will run?"

Good question. That comes after everybody is loaded into the airplane. The first good sign is when the engine starts. Sometimes when the engine has not been started recently and the air temperature is low, or the engine has been running awhile and the engine is hot, it might take a few more turns than normal to fire it up. Don't be concerned if the engine doesn't start on the first turn. Aircraft engines are built to higher standards than auto engines; you know how temperamental true artists can be.

While taxiing to the runway, the pilot checks the brakes, watches the instruments for such things as whether the compass turns when the aircraft does, and prepares to do a thorough engine *runup.* Redundancy is a key word in aviation; there are two of a great many things, just in case one malfunctions. That's true of parts on engines. There are two spark plugs for each cylinder, and two magnetos supply the electrical spark. *Magnetos* are small generators of current, independent of any other electrical system, that feed sparks to the plugs. If one fails, the other will keep the engine running.

Before beginning this final examination of the airplane and engine, the pilot will take a piece of paper, or a page in the aircraft's flight manual, that is called a *checklist.* It's not that the pilot is unfamiliar with the airplane, but the checklist makes certain nothing is overlooked or forgotten. A pilot might have flown the same airplane for years, might know each of its systems, and might know the checklist by heart. But a distraction from a passenger, a radio message, an approaching airplane either landing or taking off, or any number of interruptions could cause one or more items to be inadvertently omitted.

To prevent this, each item to be checked is recorded on the checklist and the pilot follows it step by step. A passenger in the right seat often can be helpful by reading aloud the items on the checklist and having the pilot orally confirm that they have been checked.

This might sound like a lot of work before taking to the air, and it is. But it is one of the reasons why flying is safe. Can you imagine how many fewer breakdowns of autos would occur and how many fewer accidents there would be if auto drivers would take the same precautions?

3

What's all that stuff on the panel?

Which one of these did you give me for Christmas last year?

Quickly now, list all the dials, gauges, buttons, and switches in your automobile, and tell where they are located. I'll bet you missed one or more. What about the fuses or circuit breakers? Most pilots will not be able to describe correctly the panel of their autos, but they will know the location of the instruments and switches in the airplanes they fly.

Some instructors blindfold a student, call out a particular item, and have the student touch it. This drill trains the pilot to know instinctively the location of instruments, switches, and control levers to permit rapid reaction to any emergency, and to enable the pilot to conduct an action with the hand while monitoring other instruments or air traffic.

At one point in my flying, the act of doing something without looking proved embarrassing. I had just taken off from the Baltimore/Washington International Airport, heading for Norfolk, Virginia. I was not as familiar with that particular airplane as I should have been. I was a cigarette smoker at the time. Leaving the airport area, and before reaching cruising altitude, I poked a cigarette into my mouth and pushed in the cigarette lighter on the panel of the airplane. In this particular model, the cigarette lighter is adjacent to the *fuel mixture control knob*, which, when pulled, reduces the amount of fuel going to the engine; the adjustment is necessary to achieve a proper mixture of air and fuel when the aircraft reaches higher altitudes where air is thinner. The knobs on both items are of similar shape.

Hearing the "click" indicating the lighter was hot, I reached over and without looking grabbed a knob and pulled. Yes, it was the mixture control that I pulled. One short sputter of the engine was all that was needed for me to quickly push the knob forward. Alone in the airplane, I embarrassed only myself. It was, however, another one of the many reasons for giving up smoking.

As a passenger, you need not be precise in identifying each dial, gauge, or knob. A cardinal rule for passengers is ***don't touch anything without approval from and communication with the pilot***; however, a basic understanding of the instrument panel will enable you to help the pilot, if he or she wants help. The understanding will also add to your enjoyment by eliminating apprehension and providing your own information about where you are and what the airplane is doing.

At first look, the panel might seem overwhelming. But many of the dials and gauges are familiar. They are just in an unfamiliar setting. Panels of aircraft differ depending on the preference of the owner, the kind of flying that the pilot expects to do in the aircraft, and, of course, the financial ability to "load up the panel" with all the goodies that most pilots want.

Knowing what's on the panel

Let's look at a panel of a typical single-engine airplane (Fig. 3-1). This happens to be of a Piper Cherokee 235 Pathfinder. The aircraft in which you fly might have more or fewer dials and instruments, radios, and navigation aids.

Let's begin with the very familiar ones.

In the upper left corner, item number 3 is a clock. (See, already you are familiar with at least one instrument in what you thought was a confusing array.) Next to it, item number 5, is the *airspeed indicator*. If this were your auto's dash, you'd call it the speedometer; however, it displays the speed through the air instead of telling you how fast you are moving over the ground, which as you recall from chapter 2 depends upon wind conditions. Colored lines around the airspeed indicator's dial are guides of safe ranges for lowering flaps, warnings of excessive speeds, speeds beyond which the aircraft should not be flown, and the minimum speed to maintain safe flight (Fig. 3-2).

That bank of oblong boxes in the middle is nothing but a set of radios. These tune much in the same way your car radio tunes, but often without the push-buttons for preselected frequencies, although some equipment does have the ability to store and recall frequencies with the touch of a button. The top item on the stack of radios is an *audio selector panel*, which is a box of switches to let the pilot choose which of the several radios he or she will listen to and talk over. The catch-all reference to airplane radios is *avionics*.

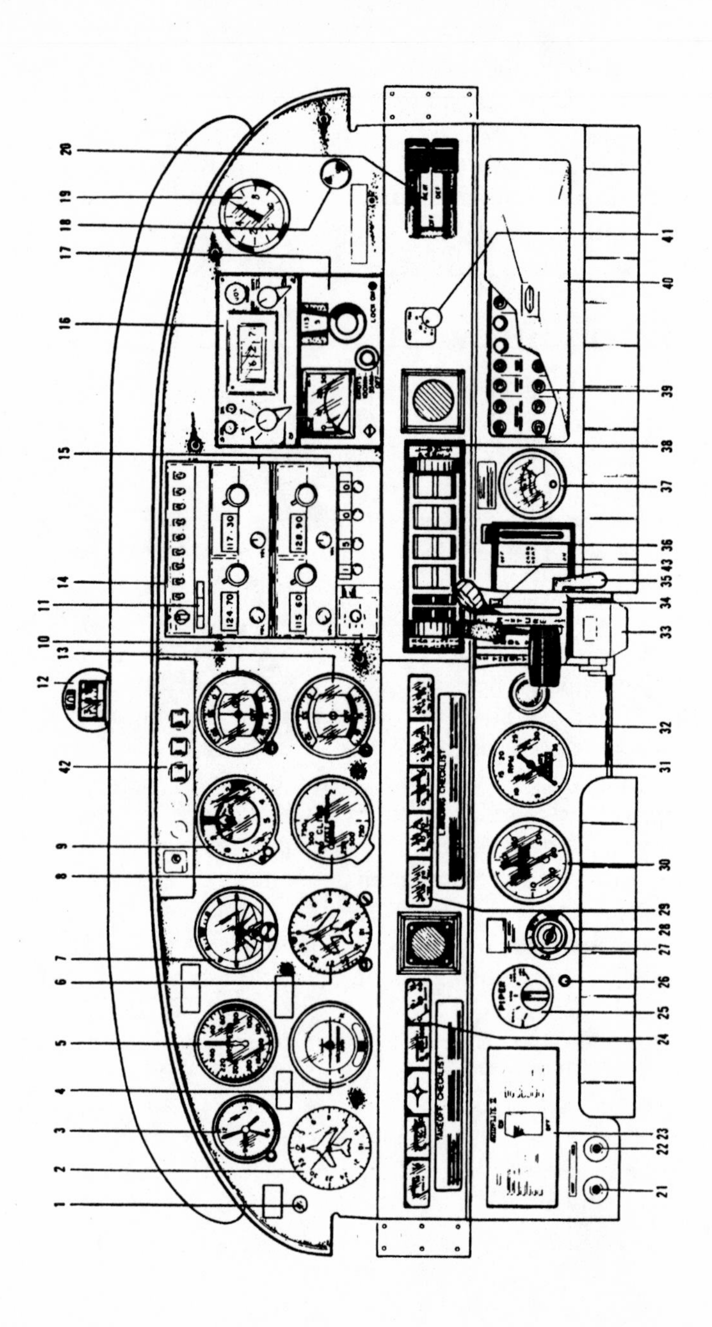
1
2
3
4
5
6
7
8
9
10
11
12
13
14
15
16
17
18
19
20
21
22
23
24
25
26
27
28
29
30
31
32
33
34
35
36
37
38
39
40
41
42
43
TAKEOFF CHECKLIST
LANDING CHECKLIST
PIPER

1. Stall warning light
2. ADF indicator
3. Clock
4. Turn & bank
5. Airspeed
6. Directional gyro
7. Gyro horizon
8. Vertical speed indicator
9. Altimeter
10. Transponder
11. Marker beacon
12. Compass
13. UHF Indicator
14. Audio selector panel
15. UHF transceivers
16. ADF radio
17. DME radio
18. Cigar lighter
19. Suction gauge
20. Heat & defroster control
21. Mike jack
22. Phone jack
23. Autopilot
24. Engine instrument cluster
25. Omni coupler
26. Nav selector switch
27. Pitch trim, ON/OFF
28. Magneto & starter switch
29. Fuel gauges
30. Manifold pressure gauge
31. Tachometer
32. Primer
33. Microphone
34. Tachometer
35. Primer
36. Microphone
37. EGT Indicator
38. Instrument panel lights
39. Circuit breaker panel
40. Circuit breaker cover
41. Optional fresh air blower control
42. Annunciator panel
43. Mixture control lock

3-1 *The instrument panel, like this illustration of a Piper Pathfinder, might seem imposing at first, but many of the items are familiar in an unfamiliar design.* Piper Aircraft Corp.

3-2
An airspeed indicator scaled to show miles per hour on the inner ring and knots on the outer ring. Arcs of different colors indicate airspeed limitations for certain actions or maneuvers.
United Instruments, Inc.

Look now at the left center area and see a line of small windows with needles; these small instruments are also familiar. The first series of windows is the cluster of engine-temperature and oil-pressure gauges, just like your car's dash. The difference is that few motorists bother to check the engine gauges until a warning light appears; a pilot pays close attention to the gauges to monitor the engine's operating health. (It's difficult to pull onto the shoulder of an airway and wait for an overheated engine to cool down.)

To the right of these gauges is the cluster of fuel gauges. (In this illustration, that square with the circle in the middle shows where the *control yoke* enters the panel; the yoke is the control device that the pilot uses to manipulate the elevators and ailerons. Nonpilots who aren't as informed about airplanes as you are now might call this the "steering wheel.") Unlike your auto, which has only one gasoline gauge, most aircraft will have at least two gauges, one for a tank in either wing.

This Piper has four gauges: two for *main tanks* and two for *auxiliary tanks.* The extra fuel capacity provides longer range. Individual gauges for each fuel tank permit the pilot to manage the fuel flow by using from various tanks to maintain a proper weight balance. Aviation gasoline weighs about 6 pounds per gallon. Imagine how the aircraft would want to tilt to one side if 20 gallons (120 pounds) were used from one wing while the other had full tanks.

To the right of this cluster are the switches for controlling the lighting system: navigation lights, landing lights, and strobe lights. At the far right are the cabin comfort levers to control the heater, defroster, and fresh outside air.

Below these comfort controls, item 39 is the circuit-breaker panel covered by item 40, just like in your car. At the top of the panel, item 12, is a *magnetic compass.* Many autos have a compass, but perhaps not quite as sensitive or expensive! Number 18 on the panel is a cigarette lighter, mercifully placed in a safe position, away from instruments with similarly shaped knobs.

Isn't it surprising how many items on the panel are just like you know in your auto? But don't let on to the pilot that you understand. It will take away the mystique with which pilots often like to surround themselves. The next time you are in a small airplane, look over the instrument panel and feel proud that many of what you once thought to be an ominous array of mystic instruments of the occult are nothing more than objects you've known most of your life.

Many items are like familiar old friends

When you see that many of the airplane's instruments are just old friends in different attire, learning about those peculiar to aircraft won't seem like such an overwhelming task. Even here, the instruments are not difficult to understand. Let's take an easy one first.

Item number 6 is a directional gyro (Fig. 3-3). This is just another compass that is more stable than is the standard compass floating in kerosene. This, like a magnetic compass, is divided into 360 degrees with the four cardinal points being north, east, south, and west. The directional gyro, however, does have a tendency to *precess* or *recess*—move to the left or right of the actual direction—and must be checked frequently against the magnetic compass and reset to agree with the magnetic compass indication.

To the right of the directional gyro, item number 8 in this sample panel, is the *vertical speed indicator* (Fig. 3-4). The fact that the word "climb" appears above the center line might have

3-3
Directional gyros are more stable than magnetic compasses.

3-4
The vertical speed indicator shows the rate of climb or descent of the aircraft. The large numbers indicate thousands of feet per minute.
United Instruments, Inc.

already given you a hint. The numbers indicate feet per minute. If the needle points to 5, for instance, that indicates the aircraft is climbing at a rate of 500 feet per minute. When the needle is pointing down from the center line, the interpretation of the instrument's display is obvious, again measured in feet per minute.

The bottom box, number 10 on this panel illustration, is a *transponder*, which emits a signal that permits ground-based radar to enhance the recognition of the aircraft on a controller's radar scope. With discrete numbers that the pilot dials into the instrument, controllers also identify that specific aircraft. Most transponders also have an altitude reporting capability, referred to as *Mode C.* Through a connection with the altimeter, the transponder reports to the controller the altitude of the aircraft, which permits air traffic control to safely separate aircraft.

The radio to the left, number 17, is a *DME,* which is an abbreviation for *distance measuring equipment.* By tuning this radio to ground-based navigation stations, it measures the time it takes to receive the signal and converts this to a gauge or numeric indicator that depicts the distance from the chosen station.

Above the DME is another radio, the ADF (*automatic direction finder*). As its name implies, this receiver picks up a signal from a selected radio station, senses the direction, and causes a needle to point directly toward the source of the transmission. Besides receiving aeronautical radio signals, this instrument usually can be tuned to commercial broadcast stations. This is helpful in locating areas that do not have aeronautical radios and very useful when the pilot wants to listen to a ballgame.

This is the appropriate time to note that the instruments most used by the pilot for navigation and operation of the airplane are directly in front of the pilot; radios and other less-frequently used instruments and switches are on the right side of the panel or elsewhere within reach of the pilot.

Looking at this cluster of dials in front of the pilot, several might appear unfamiliar. Let's look at them closely, and you will be surprised to learn that they are not that strange to you. Item number 9 on the illustrated panel is the *altimeter* (Fig. 3-5). Looking like the face of a clock, this instrument senses air pressure and converts this to an altitude readout on the dial. The small hand relates to thousands of feet (like hours on a clock) while the big hand indicates hundreds of feet (like minutes). As an example, if the small hand is between the 3 and the 4, and the big hand is on the 5, the aircraft is at 3,500 feet above sea level. No, the third hand is not an alarm. The dial is calibrated only to 10,000 feet. When the aircraft climbs above this altitude, that little nub of a hand moves to indicate that the pointer indicating thousands of feet is on its second time around.

3-5
The altimeter, when set to the current local barometric pressure, indicates height above sea level.
United Instruments, Inc.

Altimeters must be set to the proper barometric pressure. This is done prior to takeoff either by getting the pressure setting from a flight service station, an air traffic controller at the airport, or an automated radio announcement system. When these are not available, the pilot sets the elevation to the elevation of the airport before taking off. By setting the proper elevation into the instrument, the correct air pressure will be set. The pilot will make regular adjustments during a flight based upon the *altimeter setting* announced by controllers or sources of weather conditions.

Although most pilots are alert to their altitude, there was at least one reported occasion when the pilot of an airliner didn't notice the small hand indicating the altimeter was on its second time around. Making an approach by reference to instruments, the pilot descended from 14,000 feet, believing the airplane to be at only 4,000 feet. Still in the clouds, thinking that the runway should be in sight, the pilot pulled up and tried again. After the second failed attempt, the error was recognized, and the airplane descended to 4,000 feet for the proper approach in absolutely perfect visibility below a high layer of clouds.

To the left of the altimeter, item 7, is the *gyro horizon* (Fig. 3-6), usually referred to by pilots as the *artificial horizon.* This instrument shows the reference of the aircraft to, naturally, the horizon. When the aircraft is flying straight and level, the indicator (sometimes a miniature airplane design) shows straight across the center of the dial. When the aircraft turns, the marks along the outer rim of the gauge show the degree of bank. When the aircraft climbs, the indicator rises above center to show the angle of climb. Similarly, when descending, the indicator drops below the center line a distance in relation to the steepness of the descent.

3-6
The gyro horizon shows the attitude of your airplane in relation to the natural horizon. Markings on the inner ring indicate degree of bank of the aircraft.

Another instrument that shows the degree of bank is the *turn-and-slip indicator* (Fig. 3-7), positioned in the number 4 location on the illustrated panel. As the name implies, this shows the degree of bank while a ball moves to indicate the smoothness of the turn; the ball is similar to the air bubble in a carpenter's level. Remember the discussion in chapter 1 regarding rudder and ailerons used jointly to make a coordinated turn? This instrument helps the pilot to make smooth turns by using the controls to keep the ball centered while the wings are tilted, or banked. Watch the ball the next time you are flying in a small airplane. If the ball stays in the middle during the turn, a comment such as "A nice, smooth turn, Captain" would be in order. If the ball slides from side to side, the better part of discretion would be to fix your gaze outside the aircraft and avoid whistling the theme from *The High and the Mighty*.

To the left of the turn and bank indicator on this panel is the ADF indicator. The indicator points toward the station that is broadcasting on the frequency tuned on the receiver in the radio stack as previously described in this chapter. The indicator provides navigation information that the pilot can use to

3-7
The turn-and-slip indicator is also referred to as the "needle and ball." It is a combination of two instruments. The needle is gyro operated to show rate of turn, and the ball reacts to gravity and/or centrifugal force to indicate the need for directional trim.

locate some airports and to make instrument approaches to these airports in instrument weather conditions.

Examining the ADF is an appropriate point to begin learning the instruments of navigation because the ADF was the first instrument used in radio navigation. Although it still is used, modern receivers and indicators provide increased accuracy and are easier to use than the ADF.

The most common of these is the *very high frequency omnidirectional range,* which pilot lingo shortens to VOR (pronounce each letter). The two circular instruments immediately to the left of the radio stack, items number 13 on the illustrated panel, provide a readout of signals from the VOR ground-based radio facilities. The top instrument combines VOR navigation and glideslope information to provide a reference for a precision approach using an instrument landing system. The lower gauge provides only lateral directional information. How to interpret VOR indicators is fully explained in chapter 4.

The VOR is the most common means of navigation, but other systems are swiftly gaining in popularity. One is *Loran-C.* This system was intended first for marine use and has since become accepted in aviation. The U.S. Coast Guard operates and maintains Loran-C transmitters, but the Federal Aviation Administration has approved its use as an additional navigation system, including nonprecision approaches at selected airports.

Two circular dials at the bottom center, items 30 and 31 on the drawing, are instruments that indicate the amount of power being developed by the engine. Item 30 is the *manifold pres-*

sure gauge to show power settings. The manifold pressure gauge indication is used to adjust the blade angle of a *constant-speed propeller*. The second engine instrument is the *tachometer*, which is essentially the same device found in cars and trucks. The "tach" shows the *revolutions per minute* (RPM) that the engine and propeller are turning. When an airplane has a *fixed-pitch* propeller, rather than an adjustable constant-speed prop, the pilot will rely on the RPM indication to adjust the engine power.

The airplane depicted in the illustrated panel also has an *autopilot*, which will fly the airplane based upon settings given to it by the pilot plus data that is received through various navigational instruments. The switch for engaging the autopilot is at the lower left of the panel, item 23. Another switch, item 25, selects which of the VOR navigation receivers will direct the autopilot.

This sample panel contains more than what is found in many airplanes and less than what is found in others. It also depicts only one of several different layout designs. The airplane that you enjoy flying in will be different; however, the basic configuration will be the same, and the function of the respective instruments from airplane to airplane will be the same.

Other instruments

The *global positioning system* (GPS) offers precise navigational guidance that has been unseen in general-aviation airplanes. Designed originally for the United States' military, GPS is available now for worldwide navigation. *Precision approaches*, which are approaches made with vertical guidance as well as horizontal guidance, are being developed by the Federal Aviation Administration as the accuracy of the information is assured. (Chapter 4 has a complete discussion about GPS.)

Some of the other instruments that might be in the aircraft are: a *weather radar*, which searches out weather conditions in advance of the aircraft; a *flight director*, which is a complex instrument that combines several instruments into one; *exhaust temperature gauge* (EGT), which helps the pilot lean the fuel mixture to the most efficient level; landing gear *position indicator* (If the airplane has retractable gear, you might hear a pilot say "Three green" shortly before landing to verbally verify that the landing gear is extended for landing. Three green lights should be illuminated on the indicator.); and a variety of warning lights to alert the pilot to unusual conditions. The list is nearly endless as to what might be in the aircraft. For many pilots, the amount of instrumentation is limited only to how much can be crammed into a small space, or how much can be justified economically during spousal "discussions."

As a flier in the right seat, you can assist the pilot by looking up radio frequencies, tuning to the stations, checking the distances to the next radio facility checkpoint, and—extremely important—be an extra pair of eyes to double-check the pilot's reading of instruments. This is not an intrusion; it is a safety measure.

On one occasion I was flying with my family to the Atlantic Ocean shore. After crossing over the Baltimore VOR, I took up a heading that I thought was proper for the destination. It was a proper heading, but also one that would take the airplane through a prohibited area of military flight operations. Fortunately my son, not a pilot at the time, checked the navigation chart and the compass heading and gently reminded me of the error. We avoided the prohibited area and potential trouble.

A word of caution worth repeating often: Even though you understand what instruments are for and how to read them, ***do not touch or change anything without the full and complete approval of the pilot***. At the very least, touching something without permission can be very disconcerting to the pilot; at the worst, unapproved action can be dangerous.

A number of years ago, for example, I was part of a group showing European aviation personnel the ease of flying a general-aviation airplane in the United States. One particular individual insisted on checking out all the instruments in the aircraft and fiddling with them as we flew. To prevent this, I decided to put the visitor in the backseat so that nothing could be touched. Flying a Piper Aztec, we were ready to depart Fort Lauderdale, Florida, for a flight to New Orleans. My friend with the loose hands was safely strapped into a seat immediately behind my position in the left front seat: "Safe enough," I thought.

Cleared by the tower for takeoff, I added power, and the Aztec began moving down the runway. Suddenly, over my shoulders came two arms reaching for control knobs on the panel radios. Although I prefer to keep one hand on the control yoke and one on the throttles to be ready for any need to abort the takeoff, I removed my right hand from the throttles and slapped the intruding wrists. This, plus the incline of a takeoff, encouraged my frisky-handed friend to sit back in the seat. It was not the way to improve international relations, but it did improve the safety of the flight.

4

How you will find your way

I am not lost; I just don't know where we are.

Now that you have unmasked the mysteries of how an airplane flies and what those funny-looking dials, switches, and gauges are for, it's time to go someplace. That is the purpose of the airplane; it is the magic carpet that whisks you away to enchanted lands. Well, maybe they aren't all enchanted, but each does offer a welcome change of scenery and activity from the daily grind.

With no yellow line to follow on the highway, no road signs, and most of all, no gasoline stations into which you can drive and ask the attendants for help (if the ego ever permits this), how can the way be found in the emptiness of air? As they say in the Caribbean islands: "No problem, Mon."

You will find your way by pilotage, dead reckoning, and radio navigation by ground-based or satellite stations. If all else fails, the aviation equivalent of the helpful filling station attendant will help out: air traffic control radar.

As one smart crack about aviation goes, when asked how to find Miami, a pilot responded: "Just fly east to the first ocean and turn right." That's not too far from wrong for *pilotage.* You know that when looking out the window of a tall building, you are able to see landmarks at a distance. The higher the building, the farther you can see and the more landmarks you can identify. So it is with pilotage. At the airplane's cruising altitude, there are many landmarks to identify and follow: interstate highways, railroads, waterways, power lines, and the like. In the early days of aviation, a standard gasoline-station road map was the basis of navigation.

In later years, aeronautical charts became available. Pilots use two basic kinds of aeronautical charts. First, *sectional* and *world* charts are used for flight under visual flight rules; second, *en route* charts are used for flight under instrument flight rules.

This is the appropriate time to understand the difference between VFR and IFR. *Visual flight rules* (VFR) apply when meteorological conditions are better than certain minimums and pilots can see where they are going. This is usually 3 miles visibility, and the airplane must remain away from clouds. There are some exceptions to this, but you may generally consider 3 miles as ample visibility for VFR. *Instrument flight rules* apply in certain airspace all the time and in other airspace when visibility is such that reference to flight must be by instruments in the aircraft.

Under visual flight rules, pilots do not have to be in contact with any FAA traffic control facility except at or near airports that have air traffic control towers. (Only about 691 of the more than 5,500 public-use airports have control towers.) Instrument flight rules require contact with FAA air traffic control facilities and compliance with directions from controllers. Pilots might choose to fly under instrument rules even when weather and visibility are excellent, but pilots cannot fly visual flight rules when visibility is below certain minimums.

Air traffic control facilities separate traffic that is operating under instrument flight rules; however, when visibility permits visual flight rules, it is the responsibility of all pilots—even those operating within the instrument flight rules—to see and avoid other aircraft. Why don't all flights operate under instrument rules? Simple: The air traffic control system could never accommodate the volume. Approximately 80 percent of all flights operate VFR. When meteorological conditions require IFR operations, just a few of these aircraft move into "the system," which causes stress on the capacity, resulting in delays.

Rules of the road

There are "rules of the road," however, for aircraft flying VFR. It's like driving your car. There are specific lanes, granting of rights-of-way, speed limits, and other rules. Aircraft flying VFR maintain altitudes of odd thousands plus 500 feet when flying east (3,500, 5,500, 7,500, etc.) and even thousands plus 500 (4,500, 6,500, 8,500, etc.) when flying west. Aircraft under IFR rules fly at even thousands (5,000, 7,000, 9,000 etc.).

There is a speed limit below 10,000 feet. When overtaking a slower aircraft, the faster aircraft passes on the right, so the pilot sitting on the left has a clear view of traffic. Like driving, the aircraft on the right has the right-of-way when two aircraft are on a converging course. Landing aircraft have the right-of-way at airports. Aircraft arriving at and departing from airports have specific patterns to follow. So, contrary to the belief by many persons of the general public, aircraft do not just fly around without discipline any more than auto drivers move without following established laws. (That might not be such a comforting thought for anyone who drives in many congested metropolitan areas, but pilots are trained more safety-conscious.)

Navigating by pilotage

With this primer on rules, let's take a trip, using the various kinds of navigation. We will be going from Jekyll Island, Georgia, to Waycross, Georgia. We will use pilotage on the first trip. The chart we will use is the Jacksonville Sectional.

After takeoff, we see the main road leading off the island (Fig. 4-1). This is Highway 82, which we follow in a northwesterly direction. About 20 nautical miles from the airport, Highway 82 takes a turn to the west. We confirm our position by recognizing a racetrack near a small town at the point where the road makes the decided turn. Because we are in an airplane, we do not have to follow the road precisely. We can see ahead where the highway goes and "cut the corner" to again pick up the guiding route.

Just ahead we see a one-runway airport with a town beyond through which runs a railroad track. To the south of town there are three tall towers, again confirming our route. As we approach Waycross, we see a major highway with railroad tracks and power lines beside it entering the town from the southeast. To the northwest of Waycross we can see the layout of Waycross Ware County airport. Our entire trip has been made without radio contact and with navigation from what we can see from our high perch in the airplane.

Going by dead reckoning

When planning a trip by dead reckoning, we use compass headings and wind conditions. Because airplanes operate in an ocean of air and are subject to drift and airspeed changes caused by winds—just as boats are affected by currents—wind must be considered during flight planning and flight direction.

Navigation uses the direction, or *course*, measured in degrees clockwise from true north. There are 360 degrees in the compass circle. On our sectional chart, we draw a straight line

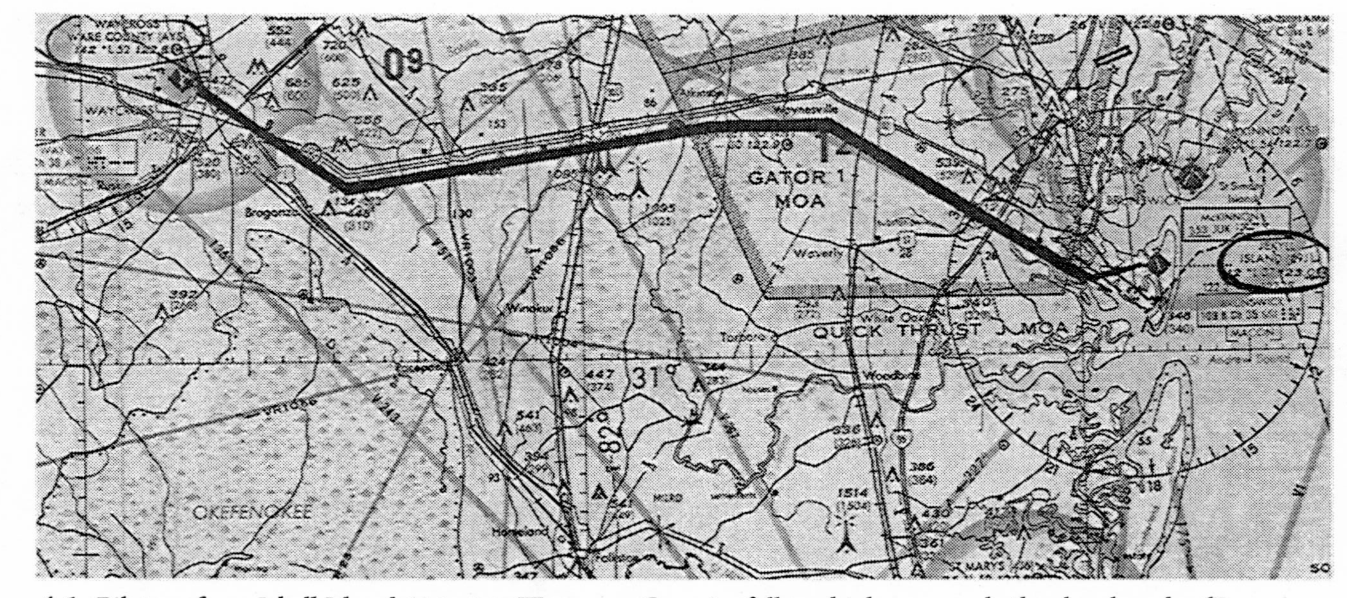

4-1 *Pilotage from Jekyll Island Airport to Waycross, Georgia, follows highways and other landmarks. (For illustrative purposes only; not for navigation.)*

from Jekyll Island Airport to Waycross (Fig. 4-2). We need to know what direction that line follows in degrees. If there were no winds, we would plan our flight for 280 degrees. This heading is found by placing a *plotter* on the chart and determining how our pencil-line route compares to the meridian. The plotter is an aviation protractor showing compass degrees and aviation chart mileage scales.

Meridians are the imaginary lines running through the North and South Poles by which positions are measured east or west of the prime meridian that runs through Greenwich, England. Because the magnetic north pole is not precisely at the geographic North Pole, this difference must be taken into account; thus, there is a variation between our *true course* and the *magnetic course.* The chart shows that the variation in this area is 4 degrees west. So, we add 4 degrees to our true course, making the magnetic course 284 degrees.

The next step is to adjust the headings to account for the winds. We must add or subtract degrees depending on the direction and speed of wind. The actual amount of *crab* that must be used to compensate for the wind is determined by calculations using an *E6B flight computer* instrument or an electronic calculator that is designed to perform aviation computations.

For this trip, the weather information we have tells us the wind is 20 knots out of 210 degrees, a southwest breeze. We know that if the wind is blowing from our left, it will be necessary to turn the airplane slightly toward the left, into the wind, to avoid being blown off track. The normal cruising speed for this airplane is 140 miles per hour; however, with a wind from the left, causing us to turn the aircraft into the wind to prevent being blown off the course track, we know that the ground speed will not be equal to the airspeed. The E6B or the calculator will determine how much crab is necessary and determine the airplane's actual groundspeed.

From calculations on the computer, we turn the nose of the airplane 8 degrees left to a heading of 276 degrees. This will compensate for the wind and keep the airplane on a magnetic course of 284 degrees. The ground speed will be reduced to 134 miles per hour. Now, as long as the winds do not change, the track of the airplane will be over the pencil line we have drawn. (If you want to have fun with your pilot, ask him

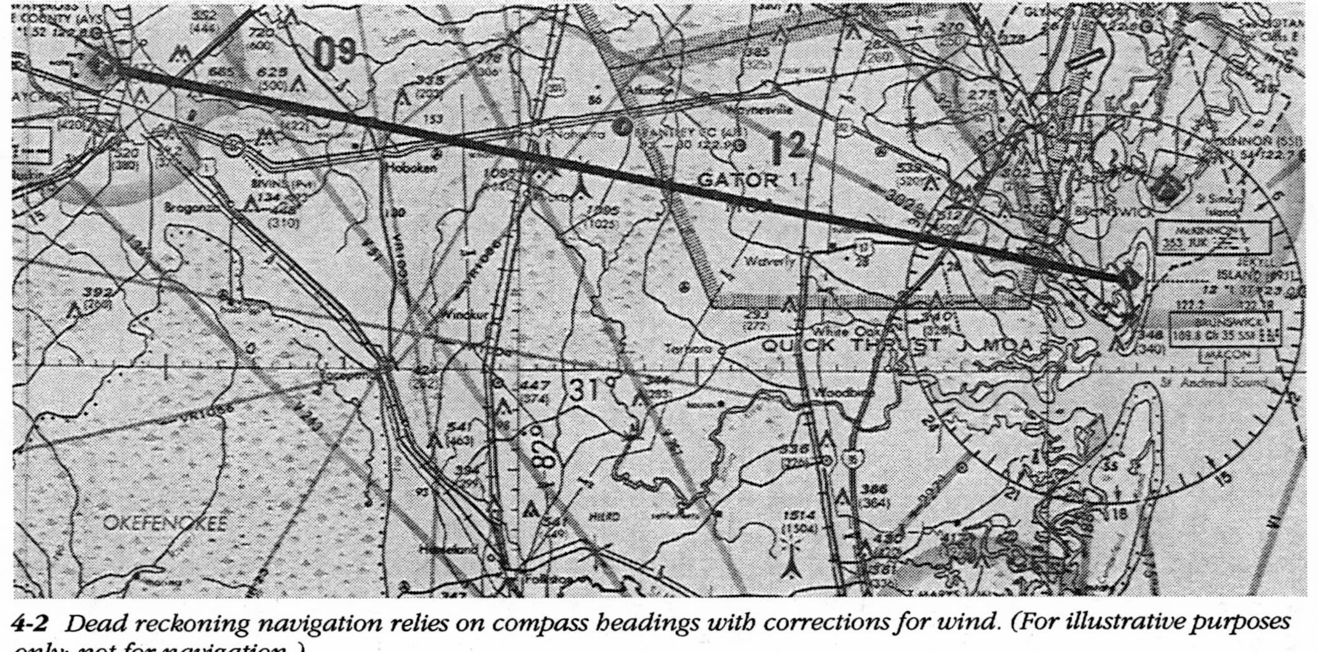

4-2 *Dead reckoning navigation relies on compass headings with corrections for wind. (For illustrative purposes only; not for navigation.)*

or her to demonstrate the use of an E6B computer to figure wind correction and ground speed. It's a rarely used art form these days, and unless the pilot is a flight instructor or just finished taking a private pilot test, he or she probably will quickly change the subject.)

To verify that our dead reckoning is correct, we check out the landmarks, or *checkpoints*, as we pass them just as with pilotage. Many of these landmarks will be the same as we saw when using pilotage, but because we are on a direct line between two points, they will be in different positions in relation to the airplane.

Pilotage and dead reckoning are useful methods of navigating, particularly in some antique or kit-made aircraft that have no electrical systems. But for most travel, navigation by radio signals is the choice of pilots. These can be by automatic direction finder (ADF), very high frequency omnidirectional range (VOR), loran, or global positioning system (GPS), which is destined to be the navigation method of the future. (Chapter 3 discusses ADF, VOR, Loran-C, and GPS radio receivers.)

The ADF is not used much in the United States for en route navigation, but is important for locating approach aids to larger airports, and often finding smaller airports that do not have more complete approach aids such as an instrument landing system (ILS). Recall from chapter 3 that the ADF points to the direction of the broadcast transmitter, so the ADF can be used for navigation by knowing the location of the station, considering wind direction and velocity, and calculating wind drift much as was done with dead reckoning. Pilots find a much simpler method in the VOR, however.

VOR navigation

A VOR is a ground-based radio facility that transmits signals to indicate 360 degrees of a compass. Each degree is considered a *radial*, and like spokes of a wheel, radiate from the center. The radial does not tell the direction of flight of an airplane, but only the position in relation to the ground station. All radials are named FROM the station. A 090-degree radial FROM a station, for example, can also be used for a 270-degree heading TO the station (270 degrees is the exact opposite of

090 degrees); therefore, the pilot must also consider the compass in the aircraft to assure correct interpretation of the VOR information.

The VOR receiving unit in the aircraft consists of: *frequency selector, omni bearing selector* (OBS), *course deviation indicator* (CDI), and a *TO/FROM indicator* (Fig. 4-3).

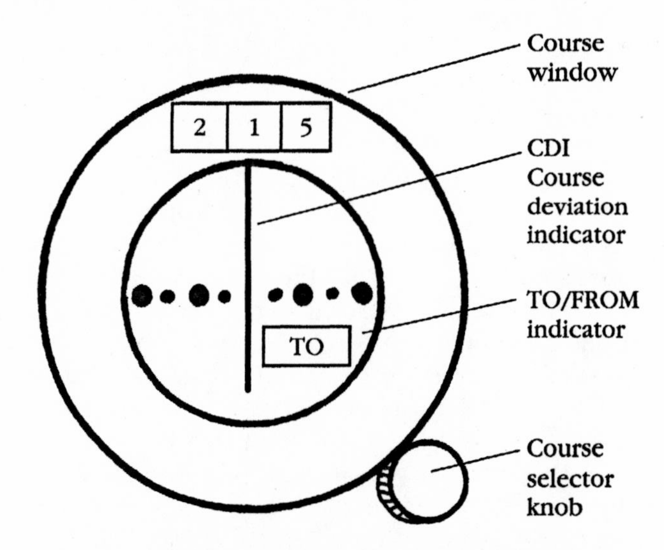

4-3 *This stylized illustration of a VOR head shows the major elements of this form of navigation. The course selector knob selects the desired compass heading, shown in the course window, with the TO/FROM window displaying a TO heading, which indicates the 215 degree heading is to the VOR ground station. The course deviation indicator (CDI) is the vertical needle in the center showing that the airplane is directly on the chosen radial.*

To select a predetermined radial, the pilot turns the knob of the OBS until the particular compass heading is in place. The CDI is a vertical needle. When the needle is in the center of the instrument, the aircraft is on the selected radial. When the position of the aircraft is off the selected radial, the needle slides to one side or the other to indicate a deviation. When off the desired radial, the pilot turns the aircraft toward the direction of the needle; if the needle is left of center, turn left to reach the desired radial; if the needle is to the right, turn the aircraft to the right. *This is true only if the aircraft compass heading is within 90 degrees of the selected radial.* If the heading and radial are more than 90 degrees apart, turns must be opposite the direction of needle deflection.

As an example, let's say the pilot wants to fly directly east of the VOR station: 090 degrees. The first step is to tune to the proper frequency of that particular VOR, make positive identification of the VOR by listening to a voice message or Morse code, and dial 090 into the OBS. (A pilot does not have to learn Morse code. Aeronautical charts show the appropriate dits and dahs for radio facility identifications.) The TO/FROM indicator will show FROM because the airplane's heading is 090 and the OBS is on 090. The face of the CDI has a center "bull's-eye" with dots on either side. When the vertical needle is centered over the bull's-eye, it indicates that the aircraft is on the radial that is indicated on the OBS selector. Two to four dots on either side of the bull's-eye indicate the degrees of needle deflection. Full needle displacement on either side indicates a deflection up to 10 degrees. The VOR indicates only the position of the airplane, not the direction it is flying. To stay on the radial, the pilot confirms a compass heading of 090 degrees and adjusts the heading to compensate for winds.

If the airplane drifts to the right of course, the needle moves to the left. If the needle is halfway from center to the left peg, the aircraft is 5 degrees to the right of course crossing the 095-degree radial. To again capture the correct heading, the airplane is turned to the left, following the needle.

Now, lets put the airplane in the same position flying in the same direction—east with a compass heading of 090 degrees—but with the OBS selector showing 270 degrees TO the station. We know from the TO indication that to reach the station we should be flying a 270-degree heading, exactly opposite our direction of flight. Again the needle moves to the left. Does this mean the radial we want is toward the left? Nope. Things are reversed. Now, instead of crossing the 095-degree radial, we are crossing the 085-radial because of *reverse sensing.* If we turn to the left now, we will be moving farther away from our desired radial. With reverse sensing, we reverse the turn and move *away from the needle.*

Centering the needle can also be helpful in locating the position of the airplane. Tune the station and rotate the bearing selector until the needle centers, then check the OBS to see on what radial this occurred and whether the indication is TO or FROM. The needle in the center indicates the airplane is somewhere along that radial emanating from the station. By making a cross reference from two VOR stations, the exact position can be determined by where these two lines cross.

Some VOR indicators also have a horizontal needle. This indicates a *glideslope* receiving capability and is used to determine position on the glidepath during a precision approach using the instrument landing system at an airport. The principle of the needles is the same. The vertical needle indicates when the airplane is left or right of the runway approach course, but the sensitivity is much higher than VOR navigation. When the horizontal needle is below the bull's-eye, the aircraft is above the glidepath; when the horizontal needle is above the bull's-eye, the aircraft is below glidepath (Fig. 4-4). In this example, the horizontal needle is above the bull's-eye and center line, indicating that the airplane is slightly below the glidepath, and the pilot should reduce the rate of descent until the needle is in the center on the horizontal line. A "perfect approach" is when the pilot keeps the two needles crossed at the bull's-eye of the instrument.

An instrument that works off certain VOR frequencies is the *distance measuring equipment* (DME); usually this is in the form of a VOR that includes the TACAN navigational systems developed for the military. The combined facility is called a *VORTAC*, which combines the abbreviation for very-high-frequency omnidirectional range and TAC from TACAN. The DME gives a readout of the actual distance the aircraft is from the ground facility. The airborne unit transmits paired pulses and the ground station responds by sending its own paired pulses back to the aircraft. The unit in the airplane observes the difference between the transmission of its signal and the reception of the return signal and calculates the distance based upon speed of radio signals, which are equal to the speed of light. Some DME equipment is capable of displaying ground speed and time to the station.

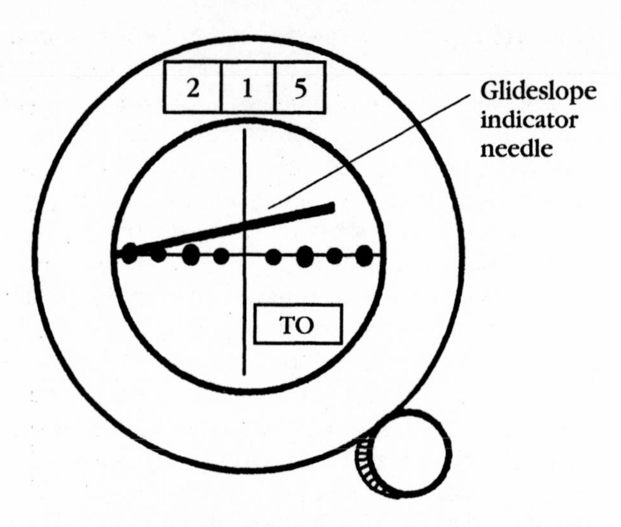

***4-4** A VOR head with glideslope. This contains all the elements depicted in **4-3**, plus the horizontal needle to display the aircraft's position in relation to the glideslope used for precision approaches. In this stylized example, the glideslope needle is above center, meaning the aircraft is slightly below glidepath.*

VORs and DMEs operate on frequencies that are limited by line-of-sight. Obstacles such as mountains, buildings, or curvature of the Earth block the transmitter's signals. Although the FAA and state aeronautics departments that have installed VORs seek the best locations, there are still blind spots, particularly at lower altitudes. Because of the curvature of the earth, the higher an aircraft is flying, the farther away the signal can be received. With no obstructions in the way, the distance that a VOR signal can be received is equal to the square root of the difference between the elevation of the ground facility and the aircraft. For example, if a VOR facility is at sea level and the airplane is at 5,500 feet, the signal can be received 74 miles away, which is the square root of 5500. (All right, if you want to get technical, it's 74.161984 miles.) At 3,500 feet, the airplane must be within 59 miles of the ground facility to pick up the signal.

As a right-seat companion, you might want to assist the pilot by changing frequencies as the airplane's course moves from one VOR to another. Perhaps you could dial in the correct radials to follow along the preplanned route. *But do not touch the radios or any other instrument without the knowledge and consent of the pilot.*

Now that we know how the VOR works, let's fly again from Jekyll Island to Waycross using VOR navigation.

While on the ground, we check the frequency of the Brunswick VOR and set it into our receiver. From the pencil line on the chart, we know that our heading will be 280 degrees FROM the Brunswick VOR. Once in the air, we maneuver the aircraft until the needle is centered with 280 showing on the OBS and the FROM flag indicating that the airplane is *outbound on the proper radial.* (Verify that the compass heading is within 90 degrees of 280; the compass ideally will be within 5 degrees either side of 280.)

Halfway to our destination we transfer VOR reception to the Waycross facility; the station is tuned and identified. The heading remains the same, but now the flag indicates TO the station. We are *inbound on the 100-degree radial.* (Check the compass.)

Two VOR receivers simplify this navigation to avoid switching back and forth; however, VOR navigation can be safely conducted by establishing the correct heading to maintain a track on the selected radial while tuning the set to the next frequency.

Through this flying under visual flight rules, we went where we chose, at the altitudes that we decided were best. A VFR flight plan might have been filed with the FAA. The safe pilot does file a VFR flight plan to assure search and rescue in the event of any difficulties.

Making the trip under instrument flight rules

Let's now make the same trip under instrument flight rules (IFR).

A flight plan is required that gives air traffic control information about the flight: the kind of airplane and registration number, the number of persons aboard, fuel aboard, estimated time of departure and time enroute, route planned, and pilot's name and address. This information is given to a flight service station by telephone, in person, or by computer. (The optional flight plan for a VFR flight is identical to the required IFR flight plan. The one exception is that a VFR pilot can plan and change the route independently, although any route change better be reported to the FSS to assist any search and rescue. An IFR pilot can plan the course, but air traffic controllers might amend the course at anytime to avoid traffic conflicts.)

Although our destination will be the same, our direction of flight will follow the *airways* (Fig. 4-5), which are prescribed routes. For our route of flight, we filed direct from the Jekyll Island Airport to the Brunswick VOR, then *Victor* 362 (all low-altitude airways are named Victor) to HABLE Intersection, then direct Waycross. HABLE Intersection is the junction of Victor 362, which is the 302-degree radial off the Brunswick VOR, and the 078-degree radial of the Waycross VOR. Because we want to go *inbound* on the 078-degree radial, we set the OBS to the reciprocal: 258 degrees.

Approaching the intersection with the VOR tuned to Waycross, the needle will show a deflection to the right. As we near the intersection, it will move toward center. (A simple way to determine that we have not passed the intersection is to mentally turn the airplane in the direction of the new heading. Then check the position of the needle in relation to this imagined heading and consider which way the airplane would have to be turned to center the needle. If the turn would be in the direction that the airplane is actually flying, the desired radial is ahead.)

The Waycross VOR is beyond the airport. We want to save some time, so we ask air traffic control for a *vector* to the airport. (Throughout the trip we have been in radio contact with ATC, and a controller has been tracking us on radar.) If the vector is approved, the controller will assign a specific compass heading that will take us directly to the airport. If the flight is under actual *instrument meteorological conditions* (IMC) (poor visibility and/or a low cloud ceiling), the vector would be to a point called an *initial approach fix* to start an instrument approach to the airport solely by reference to the cockpit instruments because the ground might not be immediately visible. (If

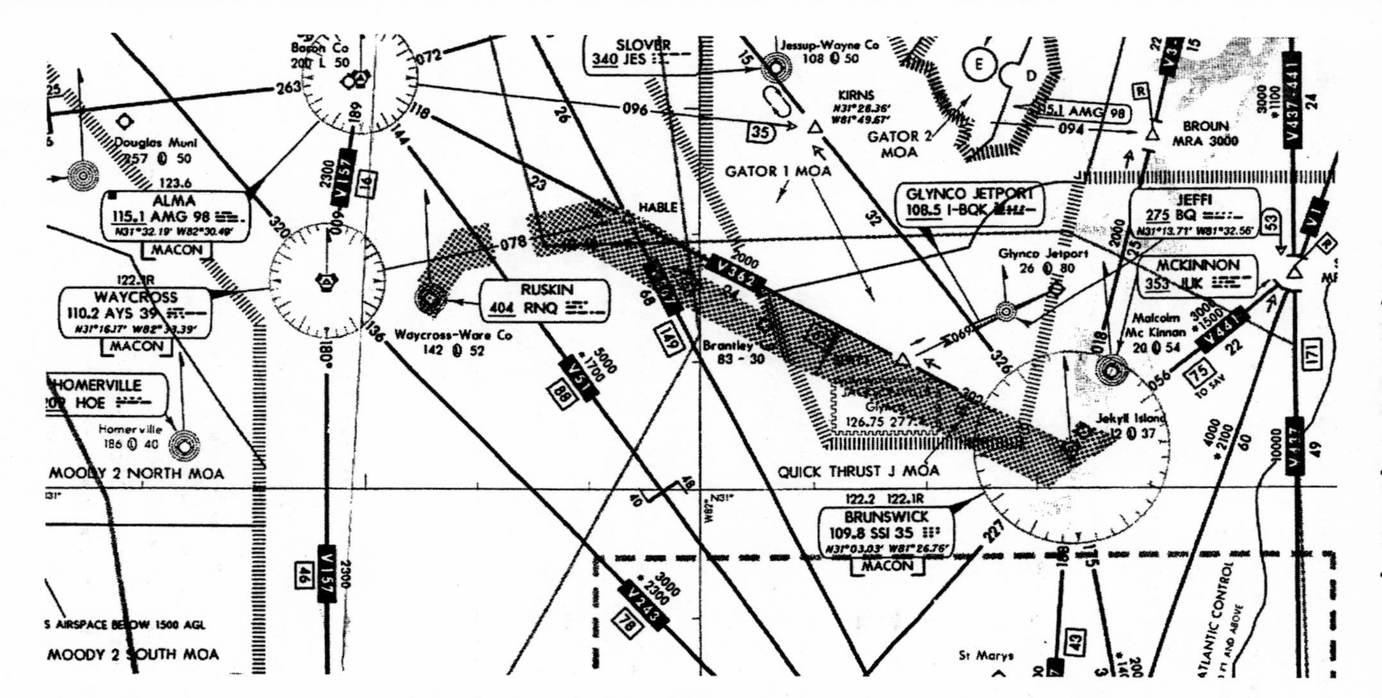

4-5 *A flight under instrument flight rules follows a more structured route along airways. (For illustrative purposes only; not for navigation.)*

the weather conditions prevent a landing and subsequent attempts to land at the primary airport, the pilot is trained to plan for that possibility and safely fly to another airport with better weather and land.)

Other ways to navigate

The VOR system is more than 40 years old but is still the most popular means of navigation for general aviation, although other systems are gaining in acceptance. *Area navigation*, called RNAV, uses the VOR but provides more flexibility. The RNAV equipment in the airplane electronically moves the position of the ground-based VOR to a location selected by the pilot. This location is called a *waypoint*. The waypoint is defined as a point that is a certain distance away from the actual location of the VOR on a specific radial. By electronically moving the position of the VOR facility, navigation can be as precise to a point without a VOR (an airport that is 45 miles from any VOR) as navigation is to the actual location of the facility.

Loran-C is a computer-based navigation system designed for marine use, but it now has some use by general aviation. Operated and maintained by the Coast Guard, loran consists of *chains* of transmitter stations. A chain is one master station and two or more secondary stations. A timing device in the receiver/processor in the airplane measures the difference between the times that signals arrive from the master and secondary stations. A computer converts this time difference to lines of position and determines a specific location based upon longitude and latitude.

Going by GPS

In February 1994, the Federal Aviation Administration approved the *global positioning system* (GPS) as a supplemental means of navigation. Approval of GPS moved aviation into a new era. There is still much to do before GPS will replace all ground-based navigation aids and before it can be used for making precision approaches, but there is no question this will be the system of the future, and it is a valuable aid at the present.

Originally designed for the military, GPS consists of 24 satellites that cover the entire world. (Lest you be concerned about national security, the satellites transmit two signals, one for civilian use that can be shut off by order of the president, and one for the military.) GPS receiver/processors receive signals from three or more satellites and process the information to form lines of position; the location where these lines cross is converted by the processor into a specific location based upon longitude, latitude, *and altitude.* From databases programmed into the receiver/processor, the instrument can depict not only location, but routes to specific locations, time and distance measurements, and other information.

GPS receivers can calculate position information within 15 meters, anywhere in the world, any time of day or night, in any kind of weather. The instrument also will tell how far away a given point on the ground—your destination airport, for instance—is from your location, the directional bearing to that point, the distance in nautical miles, the estimated time enroute, the ground track, and the true airspeed.

Additionally, at the press of a button, the receiver will identify and direct you to the nearest airports from the airplane's present location. The latter is a comforting reassurance in the event of bad weather, a sudden call of nature, or some other reason that causes the pilot to want to land at the nearest airport. No, the GPS doesn't yet do windows or take out the garbage.

GPS units might be panel-mounted or hand-held. Panel-mounted models, such as the Trimble 3100 model shown in Fig. 4-6, work off the aircraft's electrical system and usually provide the greatest amount of information. Hand-held units can clamp onto the aircraft's yoke and be removed easily to prevent theft (Fig. 4-7). These units, which can fit in the palm of a hand, take databases from various sources and offer most of the capabilities of panel-mounted units.

Like navigation charts, the databases must be updated on regular schedules. For hand-held units, the chips for updating usually may be secured by mail or parcel delivery and inserted by the owner. Most panel-mounted units require a technician to insert the updated material.

Trimble Navigation

4-6 *The global positioning system (GPS) is the navigation system of the future available now for limited use. This model of the Trimble panel-mounted receiver/processor indicates the airplane is headed to JFK Airport on a heading of 082 degrees; the distance is 988 nautical miles, which will require 3 hours and 7 minutes at 317 knots airspeed.*

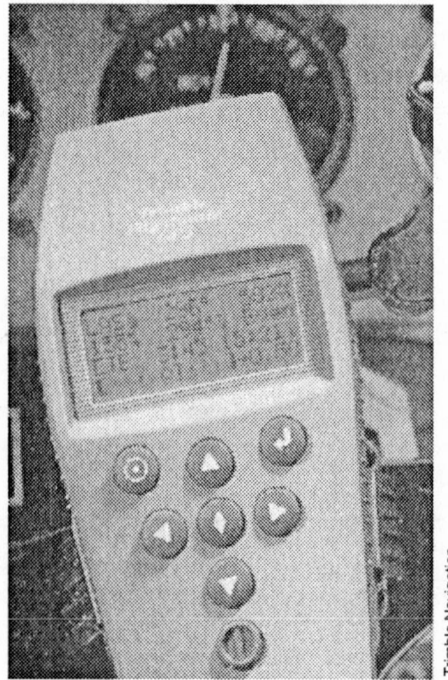

Trimble Navigation

4-7
A hand-held GPS receiver also can fit on the yoke of an airplane.

Some foreign nations are leery of turning over all air navigation to a system controlled by the United States, worrying that during times of any international disturbance, the system might not be available. However, the *International Civil Aviation Organization*, the aviation arm of the United Nations, should have a full GPS system operational sometime in the time frame of 2010–2015, about the time the U.S. satellites are expected to lose their power.

ATC is watching you

When flying IFR, the aircraft must be in contact with the air traffic control system. This is also possible, but not necessary, when flying under visual flight rules. If the workload of the controller permits, aircraft under VFR may request *flight-following service*, which means the controllers will alert the pilot to other traffic. Also, virtually everywhere in the contiguous 48 states, air traffic controllers see aircraft on their radar whether IFR or VFR. The only exceptions are a few blind spots where radar coverage is not available.

With this radar coverage, a pilot need only ask for assistance in locating a particular airport and vectors to it. Most controllers are courteous and helpful whether the flight is IFR or VFR.

Finding your way is part of the joy of flying. The right-seat companion will enjoy the flight more knowing what is happening and how the pilot is progressing along the route. It is helpful to take time to examine air navigation charts, particularly VFR sectionals. A few moments looking at a chart will reveal the secrets of reading it. Towns, roads, rivers, lakes, bridges, power lines, tall structures, and even terrain are depicted in ways that are easy to understand and interpret.

The knowledge that this brings can also be helpful to the pilot. A right-seat companion is another pair of eyes watching for correct frequencies, for radial identification, for landmarks, for airport locations, and perhaps finding new and different places to visit on the next flight.

5

Somebody down there hears you

I think you wear those just so you can't listen to me.

You never have to be out of touch with the world, although you might be a mile or two up in the air. Even pilots and passengers in antique or kit aircraft that have no electrical systems still can have radio contact by carrying small, lightweight, portable radios. Federal Aviation Regulations require two-way radios in aircraft that operate in certain airspace and at some airports. In most of the airspace and at most airports, however, radios are not required but are recommended for convenience, efficiency of flight, and safety.

The Federal Aviation Administration identifies airspace in the United States by six classes, named for the first letters of the alphabet: A,B,C,D,E, and G. These conform to the international designations that also include a classification F, which is not used in the U.S. (Fig. 5-1).

Class A airspace is that at 18,000 feet or more above sea level. Only IFR flights are permitted in Class A, and all aircraft operating there must have two-way radio capability. *Class B* airspace is that around major airports. Class B was called "terminal control area" prior to the reclassification. Clearance from FAA air traffic control and radio contact are required to enter or operate in Class B airspace. *Class C* and *Class D* airspaces are around airports that have air traffic control towers, and radio communication is required. The designation difference comes primarily from the amount of aircraft traffic that an airport handles. Radio communication is not required in *Class E* and *Class G,* unless on an instrument flight plan. These include most of the airports, heliports, and seaplane facilities, plus all the airspace below 18,000 feet, except that around the airports with traffic control towers.

There are, of course, other regulations that apply in different classes of airspace. The regulations include pilot qualification, other navigation and avionics equipment, and visibility. Let the pilot be concerned about regulations. The right-seat companion is along for the enjoyment of the trip and the rewards at the destination. If possible, pay close attention to all of the pilot's communications over the airplane's radio. You will develop a better sense of what is happening, and if you fly frequently enough you will learn to anticipate a pilot's actions and how the airplane might be maneuvered by the pilot.

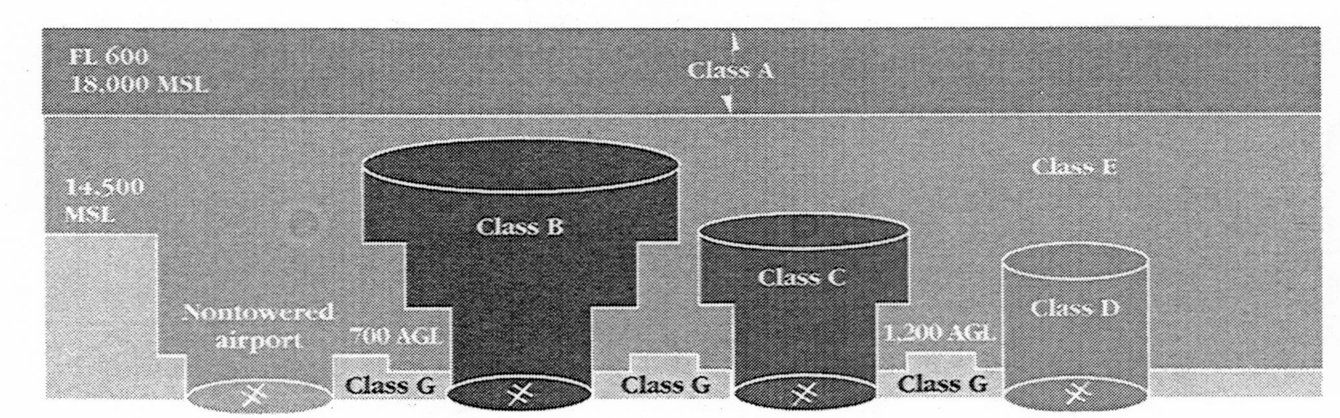

AGL - Above ground level
FL - Flight level
MSL - Mean sea level

5-1 *In 1993, the Federal Aviation Administration reclassified all airspace to conform with international definitions. The agency now identifies airspace by different classes. Flying in Classes E and G airspace does not require radio contact with any ground facility, unless on an instrument flight plan.*

If you fly often enough with a certain pilot, and the pilot is willing to teach you how to use the radio during certain phases of flight, perhaps you can become a right-seat companion specializing in communications, with the pilot's permission. This chapter will help you build a firm foundation of understanding to become comfortable with the radio.

Using the radio

The radio is a party line, not a private communications channel. Dozens of persons might be on the same frequency at the same time. That's why it is important to keep communications short and precise and to make certain the "line" is not in use before talking.

An obvious first step is to make sure the radio is tuned to the proper frequency for the type of communication. Frequencies are printed on the navigation charts. On sectional charts, the frequency for an airport control tower or an approach control for entrance into Class B, C, or D airspace is printed near the depiction of the facility (Fig. 5-2). The frequency for a flight service station serving the area is printed above the identifying box for a VOR.

Many airports do not have an air traffic control tower but maintain a *unicom* radio communications station. An airport's unicom frequency is printed near the depiction of that airport on the sectional chart; the depiction also gives field elevation and length of the longest runway (Fig. 5-3). On IFR en route charts, frequencies are listed in a table.

Listen before you talk. If more than one person is using the same frequency at the same time, both messages are garbled. Wait until the frequency is clear. Common sense also dictates that if the message you hear is a question, you can expect the frequency to be used quickly by the person responding. Wait. When the frequency is clear, it's your turn.

Sometimes a very few terminal areas will have so much radio communications that you will think a clear moment will never arise. Be ready to jump in when the opportunity presents itself. By listening carefully to the frequency, you might avoid the need to talk. As an example, if another aircraft has called a unicom frequency for wind information at a particular

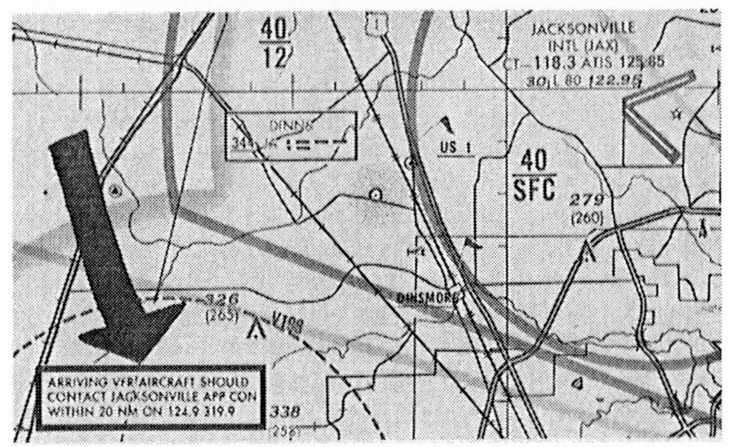

5-2 *Frequencies for contacting approach control facilities are found on sectional charts near the location of class B or C airspace. This example advises pilots to contact Jacksonville Approach Control within 20 nautical miles of the class C airspace on a frequency of 124.9. The second number, 319.9, refers to another frequency used principally by military aircraft. (For illustrative purposes only; not for navigation.)*

5-3 *Information about the airport is found near the symbol listing its location. In this example, Hazelhurst Airport is shown with an elevation of 255 feet, it has lights, the runway is 4,500 feet long, and the common frequency is 122.8. Every aeronautical map has a legend to define symbols and abbreviations. (For illustrative purposes only; not for navigation.)*

airport and you hear the answer, there is no need to repeat the question.

Plan your comments. Know what you want to say and how to say it. I have heard "ers," "ahs," "uhs," and pauses come from the most senior pilots, and I have heard crisp and precise comments come from students.

Have the microphone ready, and place it close to your lips. Key the mike for a second or two before speaking. Many pilots start talking just as they key the mike, and the first two or three words do not get broadcast. Don't be bashful or afraid that you will be embarrassed. This is not a radio show with millions of persons listening. You're just one person talking to another.

Actually, most of the comments on aeronautical radio are nothing more than repetitions of similar statements, often with only the names changed. You call because you want a clearance to take off, a clearance to land, a weather report, taxi information, a traffic advisory, or clearance to enter certain airspace.

Talking to people

On a typical flight, you will talk to one or more persons at the airport of departure, perhaps the air traffic departure controller, maybe a flight service station along the route to get a weather update, maybe an en route controller to get radar service for traffic advisories, perhaps an approach controller, then local and ground controllers at the destination airport. Let's take each of these and see what you might want to communicate to them and what you might expect.

Using the unicom

Most airports open for public use have a unicom frequency. This is used for messages and for requests for information or service. Pilots can radio ahead to arrange for refueling, to request a taxi ride into town, to alert persons to the expected arrival time, to order food to be prepared for the next leg of the trip, or for similar services. Unicom cannot be used for air traffic control; however, it can be used to provide *advisories.* At airports with air traffic control towers, one or more of the general aviation service facilities, called *fixed-base operators*, will have unicom service.

When approaching an airport that does not have an air traffic control tower, but does have a unicom, persons in the arriving aircraft will want to know wind direction and speed and which runway this favors for landing. The unicom facility can provide this information and also say whether there is any other known traffic in the area, which is the advisory.

This frequency is also used by pilots flying in the area to inform other pilots of their positions and intentions. For example, an aircraft arriving at the Frederick, Maryland, Municipal Airport would broadcast: "Frederick traffic, Piper November one, two, three, four, Golf, entering downwind, Runway 23, full stop, Frederick." This lets all other aircraft in the area know what kind of airplane it is, its position, and the pilot's intentions. Notice the airport name is used at the beginning and the end of the transmission. Hundreds of airports have the same unicom frequency, so proper identification of the location is essential.

Notice the word "Golf" is spoken for the letter G. This is to avoid confusion. Letters are spoken as the phonetic alphabet. The letter G could be misunderstood as B, P, C, D, E, T, V. (The glossary has the complete phonetic alphabet.)

Unicom is also used when departing an airport, to obtain a radio check, and to announce intentions. The radio check assures the pilot that the radios in the aircraft are in proper working order. (On three occasions I have had microphones malfunction. After the first, I have always had a spare microphone handy in the airplane.)

Talking with ATC

Air traffic controllers are professionals. Most are helpful, courteous, and willing to assist pilots in any way possible. Many also are busy. They should not be expected to take time trying to understand confusing conversations from a few careless persons in aircraft. When communicating with ATC, know your position, know what you want, and know how to ask for it.

En route controllers will provide radar *flight following service* when their workload permits, which is most of the time.

This service gives the aircraft flying under visual flight rules (VFR) information about other traffic. To request flight following, contact the appropriate facility and identify yourself. Example: "Jacksonville Center, Cessna November one, zero, eight, seven, Tango." Do not make the formal request until the controller speaks to you, which reduces frequency congestion and ensures that the controller is ready to listen to you. When the center controller acknowledges, give your position and request. "Eight seven Tango is on the three-three-zero radial, Ormond Beach, three thousand five hundred, squawking 1200, destination Albany, Georgia, requesting flight following."

Notice that after the initial identification, only the last two numbers and letter are required for subsequent communications with the same facility.

Upon approval of flight following (and probably retuning the transponder to a new code), if there is other traffic that might pose a potential conflict, the controller will radio this to you. It will be pointed out in relation to a clock. Twelve o'clock is straight ahead, three o'clock is directly to your right, and so on around the imaginary clock face. When a controller alerts you to traffic, look in that direction, and be ready to respond. If the other aircraft is sighted, immediately tell the controller you have the traffic in sight. If unable to sight the other traffic, which frequently happens for various reasons, tell the controller. "Eight seven Tango looking, negative contact." Don't leave the controller wondering if you have received the message.

Frequencies for air traffic control en route facilities are published on IFR low-altitude charts for the sector covered (Fig. 5-4). Many pilots who fly only VFR still find it helpful to have IFR charts along on the flight for the security and safety of this added information.

Talking with flight service station specialists

Along the way, you might also want to check the weather ahead and at your destination. This information is available from flight service stations. If you are on an IFR flight plan or

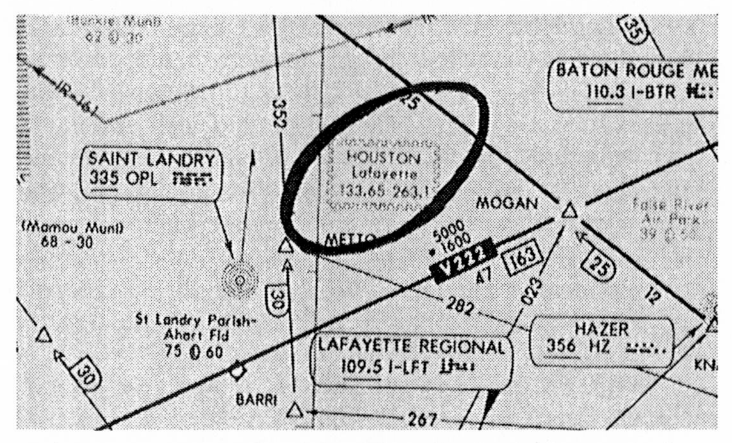

***5-4** On low-altitude en route charts, the proper frequency for the air traffic controller serving a particular segment is shown in a box located in that segment of airspace. The Lafayette Sector of Houston Center can be reached on 133.65. (For illustrative purposes only; not for navigation.)*

receiving flight following service, ask the controller for permission to leave the frequency for a few moments, then report back after completing your other radio work.

To help other pilots, you might also want to provide information about conditions through which you have flown. Maybe there is turbulence; perhaps you passed around some rainshowers; maybe winds are decreasing or increasing. Information such as this can be passed along to the flight service station specialists who, in turn, can provide it to others. Reports like this are called—imagine this—*pilot reports*; however, aviation lingo wouldn't think of using a full term that the uninitiated could understand, so the pilot community calls them *pireps* (**pi**lot **rep**orts).

Flight service station personnel are there to help by providing weather information, receiving flight plans, and closing flight plans. Flight service stations do not have radar, but many have specialized equipment to locate the position of your aircraft and direct you to the nearest airport in an emergency. "Service" is their middle name.

Talking with approach control, local control, and ground control requires the same thoughtful process of knowing what you want, how to ask for it, and how to acknowledge the transmission. To conserve time, "shorthand talk" should be used whenever possible. Consider the following: You have just landed at a busy airport and turned off the runway onto a taxiway. You could say: "Metropolis Ground Control, this is Beechcraft Bonanza November two, one, one, Charlie, we have just landed on Runway 19 and turned off onto taxiway Echo. We want to go to the XYZ general aviation facility." You have just taken up a lot of valuable radio time.

Condense your comments like this: "Metropolis Ground, one one Charlie clear of 19, going to XYZ." The ground controller knows you have just landed, he or she knows that 19 identifies a runway with associated taxiways, and general aviation service companies are known by name to the local tower personnel.

Talking on the radio is less of a problem for some right-seat companions than is listening. This is because the nonpilot often doesn't know what to expect. Usually, this can be corrected by listening to the radio more closely during flights. Or, if your pilot spouse or friend has a portable receiver, take it to the airport and listen to the communications. The words are familiar and most transmissions are repetitions of common phrases. Listen, and by knowing more about what to expect, you will find the radio language is no longer a jumble of unintelligible mumbo-jumbo.

If you do not fully understand a communication, request that it be repeated. Many controllers speak in a rapid-fire fashion. It is far better to explain on the ground why it was necessary to request information to be repeated than to have a misunderstanding that could cause a possible tragedy.

Other radio aids

Other radio aids provide valuable information without the need for anyone in the airplane to talk.

At many airports, the *automatic terminal information service* (ATIS) continuously broadcasts information that the pilot needs for landing: altimeter setting, wind direction and speed, visibility, temperature, dew point, runway(s) in use, and when appropriate any notices about the airport that a pilot should know. ATIS reports are revised as conditions change. They are identified by phonetic names that are changed in sequence as conditions change: information Alpha, Bravo, Charlie, and so on. If you hear a pilot call approach control and report "We have whiskey," it's not an invitation to a wild party. The pilot has just listened to the ATIS report and heard "information W."

When approaching an airport, listen to the ATIS on its assigned frequency, found on sectional charts in the cluster of other printed information regarding the airport. On the first call to approach control or the tower, inform them that you have "information (phonetic name)." This assures the controller that you have received the latest reports.

Another automatic service available by radio is the *automated weather observing system* (AWOS). This automated weather information gathering device collects information such as temperature, wind speed and direction, and cloud cover; the information is converted into computer-generated voice reports, which are broadcast repetitiously. Although helpful, the automated systems are considered by many pilots to be incomplete and fail to offer total area coverage. As an example, the sensors might see the area directly above the AWOS as cloudless while only a few miles away a thunderstorm could be moving toward the airport. Nevertheless, AWOS does offer information that is better than nothing if the airport is not staffed by human observers.

Still another voiceless service that aircraft radios provide is turning on runway lights at many fields for night operations. Airports that have a low volume of traffic at night usually will not have persons working at the facility during the night; nor can they afford the expense of all-night lighting for an occasional *operation.* (One operation is a takeoff or landing by an aircraft.) Clicking the microphone a designated number of times within a specified number of seconds will turn the lights on. Information about this is found in the government's airport directories. The lights remain on a set number of minutes. Most of us do not know how a touch-tone telephone works to make connections, so it's not important to learn how the mike-clicking system works except, perhaps, to the most avid engineering mind. It's important to know only that it does work, and there will be light.

You are not alone

From these basic introductions to the use of the radio, it becomes obvious that pilots and right-seat companions are never away from information and help. When contact with certain facilities might be difficult because of the quirks of radio waves, often there are other pilots ready and willing to offer assistance. Because radio frequencies are party lines, there usually are many persons listening. On more than one occasion, I have been unable to contact a facility but have found pilots ready to relay messages. As an example, when on one IFR flight, I lost communication with the sector I left and was unable to make contact with the sector ahead. Thanks to the crew of a jet airliner at a much higher altitude, my problem was relayed to air traffic control, and I received a new frequency on which contact was made.

In the event of an emergency, there is a frequency that most facilities and many aircraft monitor: 121.5 MHz. This exclusive emergency frequency is used for nothing else. When emergency help is needed, dial 121.5 and start your message with "Mayday, Mayday, Mayday." If no emergency exists, but immediate help is needed, begin your message with "Pan, Pan, Pan." Soon, on the same frequency, you should hear a welcome voice of support. An easy way to remember this emergency frequency is: "When I'm talking *one to one* (1-2-1), *I'm halfway* (.5) to safety, 121.5."

Wherever you are, wherever you are going, someplace, somebody down there does hear you. Once again it's worth repeating: When helping the pilot, make sure the communication *within the cabin* is clear, too. Don't touch *or talk* unless the pilot is aware of all your actions.

6

Waiting out the weather

The weather's going to clear any minute. In the meantime, here's something to read.

It's going to happen, so plan to make the most of it. Weather, I mean. Even the most completely equipped airplanes with the most highly skilled pilots sometimes must stay on the ground or divert to an alternate airport because of weather. Such delays or changes of plans often open new doors for enjoyment, learning, or adventure.

One of the major causes of accidents in general aviation is "get-home-itis," that pressing desire to get to the destination. The cure for this dangerous condition is recognition that it is better to miss a schedule or take an alternate means of transportation than to endanger the safety of anyone on the trip. Weather reporting is not yet an exact science. Forecasts are not always correct. This means that the most carefully planned flight might run into weather conditions that make staying on the ground the prudent choice.

An *instrument rating* for a pilot increases the opportunities to complete trips, but thunderstorms, icing, fog, low ceilings, gales, and other unpleasant weather do not respect that added rating. (A rating is added onto the pilot's license after training and testing.) Also, most general aviation airplanes are not equipped to enter *known-icing* conditions. (Known-icing means that moisture and temperature conditions are such that an airplane will accumulate ice if flown in the area. Icing is extremely dangerous and must be avoided.) Neither is an airplane structurally designed nor built to withstand the forces of severe storms. The pilot is the boss. If the decision is to stay on the ground or make an unscheduled stop, make the most of it. The smart right-seat companion will not insist on continuing a trip when the pilot has any kind of concerns. Be ready to accept a delay or change of plans. Be flexible. Often it can actually be interesting and enjoyable.

Be prepared for a stop

Take along something to do. Waiting rooms at many general aviation facilities are austere, and in some cases "austere" is a polite description. If there is any reading matter available, it usually is a selection of aviation publications that the right-seat companion would find as interesting as reading through a copy of the federal budget. Have a few of your favorite items with you to keep occupied during a brief delay.

A briefcase, tote bag, or other small container can fit in the seatback pockets or on the hat rack of the airplane and be in ready reserve for use during brief delays. Fill it with what you like to do but often don't have the time for; maybe it's a book of crossword puzzles, perhaps the latest novel that you've been wanting to read. Magazines, books on tape, and even the day's newspaper can keep you from sitting around bored with your mind focused on the delay in your travel plans.

Sometimes the delay will be more than a few hours. This opens entirely new vistas! Some of the most interesting places I have seen were visited when weathered in on cross-country flights. At Mobile I toured a battleship and a submarine; in Georgia my family and I found a unique and enjoyable seaside resort where we unexpectedly stayed overnight; in Arizona, I saw the old London Bridge; in the Carolinas I played a great golf course using rented clubs. I discovered the best homemade pie I've ever eaten at a small restaurant on the airport at Walnut Ridge, Arkansas. These are but a few of the unexpected but cherished memories of times forced by weather delays.

Find local interests

Every place you land will have some kind of local point of interest that local residents are eager for visitors to see. Anyone at the airport will gladly point them out and direct you to them. If no rental cars are available, often the manager of the airport will have a car to lend. In one town where I stayed overnight, there was no taxi service and the airport car was in use, so the airport manager called the one local police officer who pointed out the town's high points while driving me to the only motel in the area.

These longer stops are called RONs; in aviation talk, that stands for *remain overnight.* Many times the weather at your RON point will be excellent with the cause for your delay being conditions farther along your route of flight.

Every state and locality has its tourist services that provide information about what to see and do in various areas. These visitor's guides are free and offer a wide range of suggestions, and many include coupons for discount prices at some restaurants, parks, or other places of interest. Write or call these tourist bureaus, and they will send information to you. Most libraries have directories that provide addresses and telephone numbers of visitor and tourist offices.

Before starting on a trip, while your pilot is flight planning, you as the right-seat companion can look up interesting places along the way. If there are no weather delays, you might still want to just drop in to an airport to visit something that struck your fancy, a luxury not available to airline passengers.

If the weather causes an unplanned stop, the visitor's guide will have provided ideas that you can suggest as alternate stopping points. The happy advantage to travel in your general aviation airplane is that something a hundred or so miles off the direct route is only a matter of minutes away.

Some weather delays can be planned and RON points selected in advance. As an example, on one flight from Washington, D.C., to San Antonio, weather reports indicated a strong warm front with thunderstorms would be in our path. I planned a weather stop at New Orleans where we would spend the night and let the storm front pass over us while enjoying an evening of Creole food and New Orleans jazz. By midmorning the next day, the weather front had passed, and the flight to San Antonio was smooth with excellent visibility, aided, of course, by full and content passengers.

Weather can be the pilot's friend or enemy. It can also be a friend or enemy for the right-seat companion depending upon how it is accepted and dealt with.

7

You can take it with you

I didn't pack one thing more than Helen, and I don't hear Bob telling her what to leave behind.

The good news about traveling in your airplane is that you have your luggage with you at all times. The bad news is that sometimes the amount of luggage is limited.

Weight and space limits determine what can be carried in a general aviation airplane. Most airplanes have room for some luggage, but the size and weight depends on the make and model; obviously, the larger the airplane, the larger the luggage space. Some twin-engine airplanes have luggage space in wing nacelles and in the nose of the fuselage. These provide additional space, some permitting bulky golf club bags, several suitcases, garment bags, and other kinds of larger travel items to be transported with ease. This luxury, however, is not for the majority of general aviation travelers who fly in smaller, single-engine or older light-twin aircraft.

What limits the luggage

Space alone is not the determining factor. Weight is the key. Every aircraft has a gross-weight limit, whether it is a single-seat homebuilt, a four-engine jet airliner, or a weapon-toting military craft. Gross weight means the absolute limit of weight at which the aircraft is permitted to fly. This includes everything relating to the flight: the basic airplane, the instruments and radios, the fuel, the oil in the engine, the pilot and passengers, and, finally, the luggage. Sometimes a pilot might boast that "This bird can carry anything you can cram into it," but such over-gross-weight flight is not only in violation of federal regulations, but is downright dangerous.

Balance is another factor. If weight in the aircraft is not distributed properly over the center of gravity, the aircraft will have difficulty flying and, in certain conditions, might crash. Because of these factors, even though there might be room in the aircraft for additional baggage, weight and balance considerations can prevent filling the space.

The pilot has responsibility for determining just how much can be put into the airplane, the positioning of all baggage, and seating of passengers. Some adjustments can be made. The trip can be made in shorter legs, for instance, with less fuel aboard. Reducing the fuel load by 10 gallons, as an example, could increase the baggage capacity by 60 pounds. The tradeoff for this, however, is that endurance could be reduced by as much as an hour or more, so more stops must be made on longer flights. That's not all bad because the extra stops facilitate restroom breaks and might be good for restless children.

Plan in advance what to take

The time to plan baggage is long before the trip is even considered. When you are at the airport is no time to learn that the suitcases will not fit in the baggage compartment. If friends on the trip are not familiar with general aviation airplanes, let them know well in advance about the limitations.

All planning for luggage must start with the model of airplane. How big is the baggage compartment? What are its weight limits? How large is the baggage compartment door, or is baggage loaded through passenger doors? Will every seat be filled on the trip? These are questions the pilot can answer. When a right-seat companion asks the questions well in advance of the first trip, the actions will reveal a concern and knowledge that any pilot will respect.

Choosing the luggage

With the aircraft limitations known, the next step is your selection of suitcases. Usually you will find that soft luggage is better than the hardside models. Soft luggage normally weighs less than the rigid cases. Also, it is more pliable and can be pushed into the corners of the baggage area more easily, often occupying less space.

A hanging garment bag is a must. This, too, is pliable and shapes to the amount of clothing put into it. A typical garment bag can hold four to five dresses or suits, or a combination, to alleviate the need for suitcases. Many have pockets into which can be placed socks, hose, underwear, and other items. This "one stop" baggage will be appreciated at each overnight stop when loading and unloading are necessary.

Take only what you need

When packing, take only the things that you absolutely need for the activity you expect at your destination. Bulky items, such as hair dryers or blowers, take up much valuable room and weight and often are not needed at the destination. Many hotels now provide some of these personal grooming appliances in the rooms.

On the subject of toilet articles, remember that air pressure is less at higher altitudes than on the surface. This can cause some products, particularly those in aerosol dispensers, to leak. (I opened a suitcase after an extended flight at 8,500 feet and found that shaving cream had oozed from the can onto clothing.) It's best to avoid carrying any product that can leak when subjected to different air pressures.

If it is impractical to carry these items in leakproof packages, pack them separately in plastic bags or other covering where the leaking contents can do no harm to other baggage if a change in pressure sets them off.

Where you are going and what your plans are at the destination obviously dictate the wardrobe that you will pack. Common sense applies. If the trip is to a rugged festival in the country, there will be no need for a collection of suits and cocktail dresses. You would not pack hiking boots for a trip to a convention at a posh resort.

Limit the wardrobe without limiting appearance

Fashion experts recommend the "mix and match" wardrobe. For men, this means a jacket with matching trousers and slacks in a coordinated color. Shirts and ties should fit well with either the formal or casual look.

For the ladies, the same kind of combinations of skirts, blouses, and dresses, with a variety of accessories like scarves, costume jewelry, or other accent pieces can provide a selection of looks from a limited wardrobe. Magazines frequently carry articles that offer specific suggestions related to current styles.

If you are not a subscriber, they are available on newsstands and at many libraries.

Another major advantage to the mix-and-match method is that usually the coordinated colors require fewer pairs of shoes. I and any passengers try to pack no more than two pairs of shoes, even for extended stays.

When packing for children, the age determines not only their needs, but also what others can take along. Space necessary for a diaper bag, special food, and similar needs for infants can cut into space that otherwise would have been available for adults. The same commonsense approach applies for older children as for adults: Take only what they need. (Chapter 8 has more information about traveling with children.)

Comfort is the key to what to wear during the flight. Long past are the days when fliers wore special protective clothing. No one is concerned about the dirt and dust. Most general aviation airplanes have the comfort and conveniences of luxury autos: reclining seats, cool-air vents, cabin heaters, sun shields, arm rests, and carpeted floors. You will find everything that you would expect to find in any first-class method of travel.

For ease of entry and exit, women and girls might find slacks more practical than dresses. Some aircraft require high step-ups into the cabin, others might call for stepping up onto the wing, still others have a high step-up after stepping onto the wing. (The cabin-class twin or corporate jet has a small set of stairs that is typically built into the door, but these are in limited numbers in smaller airplanes.)

If more luggage is positively, absolutely necessary for your stay at the destination, pack it in separate cases and use a shipping service that can get it there overnight.

Some things should be a part of every trip, and other things should be left behind. A small, folding umbrella in the cabin, for instance, will prevent spoiling a fresh-pressed suit or dress when the unexpected rain shower appears. Cigarette lighters with flame, particularly those fueled by butane gas, should be left on the ground. The lower air pressure at altitude will make the flame flare into a bigger burst, startling the user and perhaps setting a fire. Better yet, don't smoke. Smoke in an aircraft cabin not only is annoying to others but can clog the filters of instruments, which might cause them to malfunction.

Should Fido go along?

Traveling with pets is a subject to itself. I avoid taking animals on flights, although I have flown with others who considered it routine. Animals can distract the pilot. A frisky breed increases the chances for a disturbance in the cabin. Check with your veterinarian for possible sedatives if you have any question.

Traveling with pets can bring amusing incidents. My friend Frank Kingston Smith, writer, lecturer, humorist, attorney, and aviation enthusiast known throughout the flying community, tells of one flight when he had his large dog along in his twin-engine airplane. Approaching an airport for landing, Frank had made initial contact with the tower. As he keyed the microphone for a position report, the aircraft passed a flock of birds, and his dog began barking at them. The tower controller commented on the air: "That's Frank. He's even got his dog doing his radio work!"

No two trips will be the same and no two trips will require the same baggage; however, thoughtful planning, careful preparation, and a generous amount of common sense will prove that usually you *can* take it with you—and your luggage won't wind up in Ashtabula when your airplane lands at Albany.

8

Are we there yet?

You know that latch on my pet snake's cage that you were going to fix....

Traveling in a light airplane with children can be an excellent education for them, but it can be a frustrating experience for parents, unless a good deal of preflight planning goes into the trip.

Viewing their country from the vantage point of a private airplane flying low enough to see the terrain gives the child an understanding of geography, landscapes, urban development, and farm country that no books can ever present with such dynamic force. The Mississippi River becomes a living, active waterway with steamboats and barges instead of just a blue, wiggled line on a map. The wheat fields of Kansas and the cornfields of Nebraska focus sharply on the production of food. Flying over the Painted Desert of Arizona and Monument Valley of Utah, or following the Colorado River that for millions of years has carved the Grand Canyon gives an insight into the wonders of nature that no amount of reading, or television watching, can ever provide.

Still, just sitting and staring out the window can become tiresome, and this can bring on tensions and troubles. Every age has its own problems and solutions. Some situations apply to all ages.

Careful what they eat and drink

Watch the diet before and during a flight. Avoid carbonated beverages. At altitude, the carbonization can cause gas and physical discomfort. Instead, take along fruit juices or plain water. Those great-tasting fries, greasy burgers, and colas might be fine when making a stop during an auto trip, but potential turbulence and altitude can result in queasy stomachs. Lighter foods are better.

Seat pockets should contain several "barf bags," at convenient locations, not for use only by children, but also for some adults who on occasion find that the combination of altitude, turbulence, and diet can result in an upset stomach.

Infants might be the easiest age for travel. Almost every pilot to whom I have talked reports that infants and toddlers usually go to sleep quickly after takeoff. Of course, they need proper safety-seating with safety-belt security.

The difference in air pressure at altitude can cause discomfort to ears in children or adults. Infants and children, however, are more susceptible than adults. One reason is because the Eustachian tubes in children are straight, as opposed to being curved in adults. This reduces the ability to equalize the air pressure in the inner ear to the air pressure outside the ear.

Older children and adults can take action on their own to reduce any discomfort. Yawning or swallowing usually helps. This can be aided by chewing gum or drinking fluids. If these self-help actions do not relieve the discomfort, the condition usually cures itself within a day or two after the flight with no damage to the ear drums.

Infants and children up to two-and-a-half to three years of age are unable to equalize ear pressure by using one of these helpful actions voluntarily, so adults need to be alert to ways to prevent or relieve the discomfort. Giving the infant or child fluids to drink produces the swallowing that helps to alleviate the discomfort. This should be done before the aircraft begins its descent for landing. Waiting until the child begins to cry is too late because the crying prevents sipping fluids. It is best to provide small amounts of fluids periodically, if the child is awake, and to start the swallowing process before starting the descent. This is another reason for learning to understand cockpit instruments and determine the airplane's position and estimate time of arrival without having to ask the pilot.

Not all children or adults experience this discomfort from changes in pressure. Much depends on the distribution of adenoidal and tonsillar tissues. All persons, however, can experience this discomfort if suffering from a cold. It is best not to fly with a cold.

This general information is based upon discussions with several doctors and an audiologist. Is it not intended to be applicable to every person nor is it formal medical advice. *If you have any questions or concerns, consult your pediatrician or family doctor before taking any infant or small child flying in any kind of aircraft.*

Prepare for the unexpected

Relief containers are good insurance when younger children are aboard. While pilots always recommend taking off "with full fuel tanks and empty bladders," not all children get the message. Pilot shops and many general aviation facilities offer relief containers that are inconspicuous in their design for unnoticed transport from the plane for emptying. There are models for male and female needs. The child might be embarrassed by using a relief container in a small cabin with parents, but the alternative is much more embarrassing if deplaning at a busy airport after a personal accident in flight.

Give them the right things to do

From age 3–13, children can become bored and cranky during any form of travel unless they are given something to occupy their time. Depending upon the child's age and personality, this could be anything from pipe cleaners, which can be twisted into interesting shapes of animals or other objects, to brain-testing games.

The Toy Manufacturers of America, Inc., recommend preparing a separate tote bag or knapsack for each child's toys. Bags with long handles can be slung over the back of the front seats of some aircraft for easy access by the children who usually are in the rear seats because of weight and balance considerations. What to put into those bags depends on the child's age, interests, and the length of the trip. A proper selection of items can also be helpful during those unexpected stops caused by poor weather.

Safety is the first consideration when choosing what a child should take on a flight. The Toy Manufacturers of America point out that toys are closely monitored by the federal government. Cooperation between the manufacturers and the Consumer Product Safety Commission ensure that toys are among the safest products. Still, there is no substitute for proper attention and supervision by adults. Part of this attention is selecting the proper toy for the age of the child.

Up to about the age of three years, children place almost anything they grab into their mouths. Be especially careful, the

Toy Manufacturers say, when selecting toys for children under the age of three. Avoid toys with small parts that could be swallowed or inhaled, including small balls, and avoid toys with sharp points or rough edges. Check for sturdy, well-sewn seams on stuffed animals and cloth dolls. Be certain that eyes, noses, buttons, ribbons, and other decorations are securely fastened and cannot be pulled or bitten off. The manufacturers also recommend that stuffed animals, rattles, and beads should never be attached to whatever the infant is in by any sort of string or ribbon. "No matter how harmless you might think this is," the manufacturers advise, "there is always the possibility of the cord getting attached to a button or strap of clothing or wrapped around hands, feet, or neck."

When selecting activity items for children from 3–13, think small as well as interesting. The American Automobile Association (AAA) publishes a *Travel Activity Book* for ages 4–12 that offers 144 pages of games, puzzles and other activities for times of travel. The book is available from AAA and at most bookstores.

The Toy Manufacturers of America suggest other items that are available at most toy stores: colored markers and writing pads; Magic slates; Etch-a-Sketch and Magna Doodle; View-Master with perhaps slides of the destination; magnetic playing cards; and hand-held puzzles, such as Rubic's Cube, Triamid, or Tangle.

Two or more children who are old enough to play games in the airplane would enjoy travel versions of popular books and games: Scrabble, Perfection, Shark Attack, Yahtzee, and Battleship from Milton Bradley; Travel Outburst 12 Games-in-1 or Crayon-Go-Round from Western Publishing; Travel Clue, Travel Sorry, Travel Trivial Pursuit from Parker Brothers; and Twenty Questions, Mastermind, Jeopardy, and Wheel of Fortune from Pressman.

If you want more information about selecting toys for children of all ages, the Toy Manufacturers of America offer a free booklet. It is available by sending a postcard with your name, address, and zip code to: Toy Booklet, P.O. Box 866, Madison Square Station, New York NY 10159-0866. Ask for *The TMA Guide to Toys and Play.*

Keep the airplane safe, too

Make very sure that magnetized games are kept well away from the instrument panel where they could affect the compass. Also, some electronic equipment might set up emissions that under some circumstances could affect electronic equipment in the aircraft. Be alert for small items that could become lodged underneath seats and prevent the sliding tracks from becoming firmly fastened. Additionally, any small items or unsecured items might tumble forward during unexpected turbulence or descent for landing and stick under a rudder pedal.

An activity that I used when my children were small was to offer a nickel or a quarter for every aircraft they can see while we are flying. This became particularly helpful as we approached airports without air traffic control towers. It's an incentive to have extra eyes looking for other traffic, which can add to the safety of flight and gives the youngsters something to do. I found this worked best between ages of about 8 through 15.

The great benefit from going in a general aviation airplane instead of an auto is the speed that even over longer distances permits you to answer "Yes" when asked "Are we there yet?"

9

Discovering the nation by light airplane

Make up your mind. Are we flying to the mountains or the beach?

Throw a dart at a map of the United States and just about any spot it hits will be an interesting, historic, educational, or amusing destination for a flight by light airplane. More than likely there will be an airport not too far away.

Most states have for years conducted airport development programs. These efforts, aided by interests in local communities, give the locality instant connection with air transportation for business and social benefits and provide to the air traveler access to locations unreachable so quickly by any other means.

There are more than 17,500 landing facilities in the 50 states. This includes airports, heliports, seaplane bases, and STOLports (short takeoff and landing). By far the largest number, more than 13,000, are airports for fixed-wing aircraft. Some of these are private and not open to the public, but more than 5,500 welcome transient aircraft. Many of the private strips may be used with prior approval of the owner.

Usually pilots prefer to have a destination for a flight, rather than just "boring holes in the sky" to satisfy their enjoyment of flying. For the right-seat companion and other non-pilot family members, experiencing pleasant times at favorite and familiar haunts, or seeing new places, makes the trip in the airplane more than something to humor the person in the family who is the pilot.

Describing every place that is worth visiting by airplane would fill volumes, and there are volumes that report this. What follows is but a hint at the unlimited variety of destinations awaiting you in your airplane. I have visited many of these; others have been recommended by fellow pilots or local tourist bureaus. AOPA's *Aviation USA* is an excellent source of information about airports near these destinations or any destination that your dart finds.

East

Starting in New England, land at one of the airports on the Maine coast for an excellent lobster dinner. Try Eastport where you can also visit Campebello Roosevelt International Park. If you don't want to fly that far north for a Sunday brunch, drop in at Bar Harbor, Rockland, Portland, or any of the interesting coastal towns. At almost any stop you will find delicious foods and picturesque communities.

Maine is more than rugged seacoast. Its inland regions, which comprise Maine's lakes and mountains, reveal beautiful and recreationally rich areas for boating, fishing, hiking, and especially challenging golf courses in the hilly terrain. In the heart of the state, the Kennebec and Moose River valleys offer whitewater rafting. A fall foliage flight over any part of the New England area is a breathtaking experience.

Winter sports enthusiasts will find New Hampshire a definite destination for their flying. There are Alpine (downhill) and Nordic (cross-country) ski areas and opportunities for snowmobiling. Many special events are scheduled during the ski season that might include ski and snowboard racing or inn-to-inn ski touring.

Special events abound in most states throughout the year. In Massachusetts, for instance, Boston hosts a week-long Harborfest around the Fourth of July, which is a festival that includes more than 125 events including concerts, cruises, historical military drills, and fireworks. To please the palate, participate in the annual Chowderfest during this time, and cast your ballot for the "Best Chowda" served in the city. The Boston Pops conducts its Fourth of July concert on the Charles River Esplanade. If you want to avoid Boston's Logan International Airport, land at Hanscom Field at Bedford, Beverly Municipal, or Norwood Municipal.

Any family member interested in museums and history will find that Connecticut offers many options. The Groton-New London Airport provides access to the history and recognition of the submarine service. Here you can board the first nuclear-powered submarine, the *Nautilus*, stroll displays that recount the exploits of the submarine service, view through periscopes, and see a working submarine control room.

Nearby Mystic Seaport takes you back in time to nineteenth-century sailing. There you may board authentic sailing ships, see a maritime village with historic homes, and watch craftspeople at work.

Aviation buffs will want to stop at Bradley International Airport for a visit to the largest indoor display of historic aircraft in the Northeast. The New England Air Museum displays more than 70 aircraft ranging from classic private and commercial airplanes, helicopters, and gliders, to World War II fighters and bombers (Fig. 9-1).

New England Air Museum

9-1 *The 1930 Thompson Trophy winner, The Laird Solution, is one of the historic aircraft on display at the New England Air Museum.*

The Amish country of Pennsylvania offers quick-trip or extended-stay destinations that provide outlet stores, quaint shops, and an unusual selection of crafts. Reading or Lancaster Airports are good stopping points.

Anyone who has an interest in the Civil War—this should include every student—will find the private airplane taking them back in time to visit the sites of the most prominent battles. There are many ways to do this. If your home base and aircraft permit frequent trips to the battlefield areas, these sites can be visited on a series of weekends. Gettysburg Airport is just 2 miles west of the city. The Manassas Municipal Airport is only 10 miles from the Manassas Battlefield Park.

A more complete way to relive the Civil War is to fly to Frederick, Maryland, and make this your base for a series of daily tours. Frederick is the location of the Barbara Fritchie Museum, which is the reconstructed home of the war heroine made famous in the poem by John Greenleaf Whittier. "Shoot if you must this old gray head, but spare your country's flag she said" is Whittier's account of Fritchie's call from her upstairs window to General Thomas "Stonewall" Jackson.

Just south of Frederick is the site of one of the least known, but vitally important battles of the war. The Battle of the Monocacy pitted a bunch of Northern recruits, still in training, against the forces of Confederate General Jubal Early. This raid of Maryland communities was designed to draw Union Forces away from Robert E. Lee's army at Petersburg, Virginia. General Grant responded by sending a 5,000-man division northward. Until they arrived, the only force between Early's advancing Confederates and the nation's capital of Washington, D.C., was the ragtag recruits of Major-General Lew Wallace, most of whom had never seen any battle. (Wallace later became better known as the author of *Ben Hur.*)

These untrained soldiers rushed from their training center near Baltimore to engage Early's forces near the Monocacy River. Although the 2,300 Federals fought fiercely, they were no match for the nearly 18,000 Confederates. By late afternoon, Wallace's forces retreated toward Baltimore leaving 1,600 dead, wounded, or captured. They lost the battle, but the delaying tactics gave time for the reinforcements to arrive and establish a line of defense just north of Washington. The capital was saved from capture.

Antietam Battlefield is just 25 miles west of Frederick. Here, on September 17, 1862, more than 23,000 men from both sides were killed, wounded, captured, or missing in what was the bloodiest single day of the Civil War.

Near Antietam is Harper's Ferry National Historical Park, the site of John Brown's raid, which attempted to capture guns with which to arm the slaves before the Civil War began. Gettysburg is just 34 miles north of Frederick, and Manassas is less than a 2-hour drive away. All of these battlefield sites are administered by the National Park Service and include displays. The largest battlefields have road tour routes with markers that explain all the actions of the events. Often Civil War buffs reenact the battles. Check with the states' tourist offices for dates and additional information.

While at Frederick, aviation enthusiasts can visit the headquarters of Aircraft Owners and Pilots Association, which is located on Frederick Municipal Airport.

For an even earlier touch of America, land at the Williamsburg-Jamestown Airport in the historical triangle of southeastern Virginia. Here the Colonial highway connects Jamestown, Yorktown, and Williamsburg.

Colonial Williamsburg lives today as it did nearly three centuries ago when it was the social, cultural, and political capital of England's largest colony in the New World. Costumed hosts and hostesses interpret this important chapter in America's history for today's visitors.

Just 12 miles east of this living monument to history is Yorktown where the United States independence was won in the final battle of the Revolutionary War, after which General Cornwallis surrendered to General Washington. You can relive this proud period in the nation's history through multimedia exhibits and films. The battlefield is within easy walking distance from the museum.

In 1607, Captain John Smith and 103 fellow settlers stepped ashore on Jamestown Island, only 8 miles west of Williamsburg, to begin the first permanent settlement in the New World. The remains of a glass-blowing factory are preserved, and there are a reconstructed James Fort, an Indian village, and full-size replicas of the three small sailing ships on which these settlers traveled from England.

If all this saturation in history becomes too much for the children, or adults, just 3 miles from Williamsburg is Busch Gardens, The Old Country. From its breathtaking rides to its glittering shows, Busch Gardens captures the delights of Europe's feasts and sights.

The Carolinas offer unlimited recreational destinations. Some of the finest golf courses in the East are in the Carolinas. Many resorts offer golf and tennis packages along with plush accommodations and gourmet dining.

One place in the Carolinas that every pilot would want to touch the airplane's wheels is the Kill Devil Hills Airport at Kitty Hawk. This 3,000-foot runway is near the spot where Orville and Wilbur Wright made the first flight. The two shacks that served as a hangar and living quarters for the brothers still stand (Fig. 9-2), and visitors can walk in the sand along the paths of the four flights of that day in 1903. In the Wright Brothers' Museum, expert guides explain the beginnings of

9-2 The two sheds used by the Wright Brothers when they mastered controlled flight still stand within walking distance of the Kill Devil Hills Airport.

flight as the Wrights developed it. There is no overnight aircraft parking at the Kill Devil Hills Airport, but just minutes away is the Dare County Airport where you can bed down the bird while you enjoy the Outer Banks.

Savannah, Georgia has been called "the most beautiful city in North America" by Le Monde in Paris, France, and Savannah is listed as one of the 10 best walking cities in the United States by *Walking* magazine. This provides an ideal combination to enjoy the beauty of the city at your own leisurely pace.

All along the Atlantic Coast the light airplane traveler will find whatever recreation is wanted. As an example, Jekyll Island today draws on its historic heritage of fifteenth-century gold-seekers, seventeenth-century pirates, and its eighteenth-century plantation era to provide a combination of recreation ranging from four challenging golf courses, to deep-sea fishing, to peddling along 20 miles of paved and peaceful bike paths (Fig. 9-3).

Nature lovers will want to stop at the Waycross-Ware County Airport, which is the northern gateway area to the Okefenokee Swamp, the largest national wildlife refuge in the eastern United States. This vast peat bog was once a part of the ocean floor and now is a 600-square-mile "lost world" made up

Jekyll Island Authority

9-3 *More than 20 miles of paved bicycle paths are just one of the attractions at Jekyll Island.*

of freshwater prairies for marshes and huge cypress trees (Fig. 9-4). Bird, mammal, reptile, amphibian, and fish life is abundant and varied. Okefenokee Swamp Park, just south of Waycross, offers exhibits, trails, boat tours, serpentarium, and a wildlife observatory.

Your light airplane can take you to the less obvious destinations so often missed by ground-bound travelers who find only the most well-known destinations. The 5,000-foot runway at Sandersville, Georgia, is the route to seeing beautiful southern homes and the old jail in which Aaron Burr was held for a night.

The Telfair-Wheeler Airport at McRae, Georgia, provides a 4,000-foot runway, which is just across Highway 411 from the Pete Phillips Lodge where you can enjoy either a fine meal at the Fairway Restaurant, overlooking the sixth green, or a longer stay for golf and activities at Little Ocmulgee State Park where there is fishing and boating. While adults are playing the 18-hole championship golf course, the more timid or younger

Jimmy Walker

9-4 The gnarled swamps of Okefenokee invite the adventurous.

set can go the rounds playing miniature golf on a replica of the big course. The Sky Valley Resort is an all-year resort with the only snow skiing in Georgia. The nearest airport is near Franklin, North Carolina.

Flying over Florida, one is rarely out of sight of an airport. Stop at almost any one and there will probably be something of interest, a special restaurant, a beach, resort, or other point for rest and fun. Of course, in the middle is Orlando and DisneyWorld. Orlando International Airport, Orlando Executive, Central Florida Regional, or Kissimmee are potential stops for visiting DisneyWorld, Sea World, Universal Studios Park, or other attractions in the area.

Many general aviation pilots don't fly farther south than Miami, and they are missing some of the best travel places in the state—the Florida Keys. All along the string of islands you will find excellent restaurants, an assortment of water activities, and unusual attractions. Humphrey Bogart fans will remember the classic motion picture *Key Largo* and can visit the building and dock where much of this picture was filmed.

Also at Key Largo is Pennekamp Coral Reef State Park, the only living coral reef in the continental United States. Glass-bottom boats, scuba gear, or snorkel gear take you to this fully protected marine sanctuary where you can become friends with more than 500 species of fish, or explore the wrecks of ships that have fallen prey to the reef, including a Spanish galleon and a World War II freighter sunk by a German U-boat.

There are a number of private airports on the Florida Keys, some of which permit transient planes to land. Your best move, however, is to land at Homestead General Aviation Airport. Or fly over the islands to Marathon, Florida, where you will find a 5,000-foot runway. This will put you about halfway to Key West, and with a rental car you can explore both directions on the Florida Keys from this base.

Central

The ultimate in flight is zooming into space, and Huntsville, Alabama, provides an opportunity to experience the sights, sounds, and gravity forces of real space flight at the U.S. Space and Rocket Center. Here you will find the nation's only full-size shuttle exhibit open to the public and a collection of rockets developed in Huntsville.

It's only a short flight from space to the sea and the U.S.S. *Alabama* Battleship Memorial Park at Mobile. The *Alabama* won nine battle stars in World War II. You can "man" the anti-aircraft guns, which shot down 22 enemy aircraft, and see the 16-inch guns and their 2,700-pound projectiles.

Kentucky has the Derby on the first Saturday in May, but there are year-round attractions worth flying to in this state that stretches from the Blue Ridge Mountains to the Mississippi River. Land at the Bowling Green-Warren County Regional Airport, for instance, and you are just minutes away from one of the world's most unusual caves. Open every day of the year except Christmas, Mammoth Cave has the world's longest network of cavern corridors, covering more than 300 miles in length. These colorful avenues provide startling views of stalactite and stalagmite formations. On the half-day tour, which takes about 4 1/2 hours, you can experience dining in the Snowball Room, 267 feet underground.

Louisiana has only seven airports where passengers on scheduled airlines may go, but in your general aviation airplane, Louisiana welcomes you at 80 airports. One of them, Jennings, arranges for you to taxi your plane right up beside your Holiday Inn room (Fig. 9-5). Besides its local attractions, such as Cajun food, wildlife refuges, and Zigler Museum containing one of the South's finest art collections, Jennings and Jefferson Davis Parish often play host to fly-in groups.

Burt Tietje

9-5 *Taxi your airplane right up to the door of your Holiday Inn at Jennings, Georgia.*

Land at the Lafayette Regional Airport and you are in the center of a fascinating culture—Cajun Country. You will find dishes like gumbo, boudin, red beans and rice, and jambalaya. Just west of Lafayette, visit Braux Bridge during the Crawfish Festival. On the historic Bayou Teche, this town takes its crawfish eating seriously with the festival bringing out costumed characters and in-the-street fun.

Also in the area, see Chretien Point, reputedly a hangout for smugglers, including Lafitte and his lieutenants (Fig. 9-6). A few miles south, see the plantation home of Emmeline Labiche, immortalized as the Evangeline in the poem by Henry Wadsworth Longfellow. The Acadian House, built in 1760, is

9-6 *Chretien Point, formerly a hangout for smugglers, is one Cajun attraction in Louisiana.*

now a museum containing furnishings and other items depicting the way of life of the early Acadian settlers.

In the classic story *The Wizard of Oz*, Dorothy clicked the heels of the magic red shoes three times and she was back in Kansas. Your airplane requires a little more than three clicks to whisk you there, but it offers you nearly as much magic in the variety of activities to which it can take you. Besides boasting Wichita as "The Air Capital of the World," where Beechcraft, Lear, Cessna, and Boeing build aircraft for the world, Kansas offers the traveler a mixture of old west and modern heartland of America.

To gain a touch of old Sweden, pick the airport at either Salina or McPherson for a landing; about halfway between the two is the town of Lindsborg. Known as "Little Sweden, USA," Lindsborg boasts a gallery honoring the Swedish-American painter Birger Sandzen, and the Swedish Pavilion, which was taken from Sweden to St. Louis for the 1904 World's Fair. A treat of the Swedish delicacies, alone, is worth the stop.

Land at Dodge City and you can find at the site of the old Boot Hill cemetery a re-created Dodge City of the 1870s, complete with Miss Kitty and the Long Branch Saloon. Attractions in the summer include medicine shows, staged gun fights, and stagecoach rides. While Dodge City, once known as the wickedest little city in America, recalls the desperadoes of the last century, other locations in that state testify to the pioneer spirit and struggles for safety and freedom. Visit Fort Leavenworth or Fort Scott—each has an airport—and see how the west was protected. The entire Fort Leavenworth facility is a national historic landmark. A booklet for a self-guided tour is available at the post's frontier army museum. Fort Scott will take you back to the middle of the past century as you tour America's only authentically restored military post of that era.

For a touch of later history, land at the Abilene Municipal Airport and see the Eisenhower Center that honors the five-star general who led the Allies to victory in Europe and later became president.

For a look at the totally different lifestyle of another president, fly north to Dickinson, North Dakota, then drive into the Theodore Roosevelt National Park. It is said that here is where Roosevelt found his "perfect freedom" riding through the savagely beautiful Bad Lands. To sample this cowboy life, saddle up and ride into these same rugged areas where the buffalo still roam and cross-country trails take you to places like a petrified forest.

In South Dakota, a visit to Mount Rushmore, where the faces of four presidents are carved into the side of a Black Hills mountain, is a must for every visitor. In your own airplane, you can take a flight over the mammoth sculpture. But don't limit your visit to this area. Fly northwest to the Spearfish Airport, rent a car, and enjoy some of the country's most beautiful scenery as U.S. Highway 14A meanders through 19-mile Spearfish Canyon. Then drive to Deadwood and Lead. Walk the streets where Wild Bill Hickok and Calamity Jane trod. On Deadwood's historic Main Street, amble over to Saloon No. 10, where Wild Bill was murdered holding the deadman's hand of aces and eights (Fig. 9-7).

In the northeast corner of Oklahoma you will find Shangri-La—literally. On Grand Lake of the Cherokees, the Shangri-La resort offers excellent golf, boating, and other water activities, plus fine food. An excellent strip with clear approaches immediately adjacent to the resort had been closed for several years but should be reopened by the time this book is published. You might want to land at Grove Municipal's 3,398-foot paved

9-7 The west is alive in Deadwood, South Dakota, where stagecoach rides carry you along the streets made famous by Wild Bill Hickok and Calamity Jane.

strip, which is 16 miles away but offers courtesy transportation to Shangri-La.

West

Flying the Rockies brings views of the world that are wondrous and awe-inspiring. As you pass certain areas, the sights almost cry out for a landing to explore the untouched, uncrowded beauty of canyons, streams, forests, and lakes. Some of Idaho's richest recreational areas for flipping a fly in a calm pool or casting into a whitewater stream are accessible best by airplane. If the pilot has not flown into these before, however, it's wise to get a thorough briefing from experienced local pilots. Many of these airports are hidden deep in canyons that must be approached in a particular way for safety.

The Old West wasn't all cowboys. As the Indian communities in Idaho, New Mexico, and Arizona attest, the culture of the Native Americans still lives on the reservations. Just north of Pocatello, the ruts of the old Oregon Trail lead past the site of the original Fort Hall, which was a stopping point for pioneers going west. Visit a replica of the fort in Pocatello.

A stop at Santa Fe, New Mexico, puts you in the highest-elevation state capital in the United States. Head for the plaza, which is the hub of the historic district where the Native Americans sell their wares of turquoise and silver.

The sight of the Teton Range of mountains makes a flight to the Jackson Hole Airport in Wyoming worth even a lengthy travel time. This is also the entrance airport to Yellowstone National Park. Visit this one from late spring through early fall.

The vast expanse of Montana offers the general aviation traveler a host of interesting destinations that the ground or airline traveler will miss. Fort Benton, on the Missouri River, contains historical sites that date back to the Lewis and Clark expedition in the early 1800s. Fort Benton's airport has a 3,400-foot paved strip.

At Libby, Montana, you can land at an airport that boasts a starring role in a motion picture. This field was the location for filming the movie *Always.* The airspace over Glasgow is used by Boeing for flight testing. Nearby Fort Peck Lake offers great fishing, water recreation, and wildlife viewing.

Arizona, like most states, could have a complete book detailing interesting places to visit. Just flying over the state is a fascinating experience. From the Grand Canyon in the north to the border towns of Yuma and Douglas, Arizona offers a variety of stops ranging from nature's beauty to a host of outdoor activities.

Along Lake Havasu, many resorts and marinas beckon the persons looking for a different vacation or weekend. Bullhead City, for instance, is a mecca for lovers of water activities. Houseboat rentals provide living quarters while trying to catch a record-size striped bass. Waterskiing, jetskiing, windsurfing, and scuba diving are available, and just across the river is the nightlife and gambling of Laughlin, the third largest area in Nevada for gambling receipts.

Farther south along the Colorado River, Havasu City mixes the old and the new. This community, only 30 years old, boasts an attraction that dates back more than a century: London Bridge. During the mid-1960s, the city of London, England, discovered that the unstable clay of the Thames River was making the bridge, which had served the city well since 1831, unstable. Like the children's song foretold, London Bridge was really falling down. The founder of Havasu City bought the bridge. It took three years to dismantle the structure, mark each of the 10,000 pieces, and reconstruct the bridge, with a concrete core for stability, over Bridgewater Channel (Fig. 9-8).

Lake Havasu Area Visitor & Convention Bureau

9-8 *London Bridge, at Havasu City, spans a mecca for water-sports enthusiasts.*

On the eastern side of the state, land at the Holbrook Airport, just 3 miles north of what, in the 1880s and 1890s, was called "the town too tough for women and children." This history can be relived with a walking tour of the "Bucket of Blood" saloon and the "Blevins House" to recall one of the most famous shootouts in the state that closed the chapter on cattle and sheep range wars of northern Arizona. Just a few miles out of town are the Painted Desert and the Petrified Forest of trees that fell more than 200 million years ago.

With more than 250 airports, California offers a selection of destinations that can take the people in an airplane from sunbathing to snowskiing on the same day or from a quick lunch of buffalo burgers at a restaurant on Catalina Island to a dinner in Sacramento. Make a stop at the Nut Tree, near Vacaville, where a miniature train takes you from your airplane to the restaurant. At the Nut Tree, you will find the largest outlet center in the western United States with 130 stores.

For a unique experience, ask your pilot to land at the Thermal Airport, and watch the altimeter unwind. This airport is 117 feet below sea level so when the airplane's altimeter shows zero, the airplane is still more than 100 feet from touchdown.

To select any one place in the state would not do California justice, but one not to miss is a landing at Columbia Airport. Less than a dozen miles away from this historic mining town is Angels Camp, home of the Jumping Frog Contest made famous by Mark Twain. West of Columbia is Yosemite National Park. Nearby you will also find skiing at Dodge Ridge, swimming and boating at Pinecrest Lake, and year-round fishing at New Malones Lake and Dam.

Many of the places of lodging in the area offer year-round activities. Long Barn Lodge, for instance, is just 21 miles from the Columbia Airport and is a family-oriented retreat. You have your choice of lodge rooms or cabins with kitchen facilities. For winter, there is an indoor ice-skating rink, and summer finds visitors lounging around a heated pool or taking hikes or horseback trips among the big pine trees of the Sierras.

Go for it

Columbia is the headquarters of a unique travel plan that is available only for pilots and their passengers. Bill and Fritz Maasberg operate the Pilot's International Bed-and-Breakfast Fly-Inn Club. An initial $20 fee and an annual $12.50 dues buys a subscription to Maasberg's listing of places, run by other pilots, that offer bed-and-breakfast accommodations. Listings include information about the airport and points of interest nearby.

This sampling of what is available by general aviation should make you want to get into the airplane and begin enjoying the life that awaits pilots and their passengers. But accept this word of warning: Travel by general aviation can be habit-forming. Perhaps even more alarming, learning from the right-seat what aviation has to offer can lead to abandoning that position for a move into the pilot's seat. After that, you can invite other right-seat companions to lose their white knuckles.

Glossary

The pilot shot an ILS, but at DH was told RVR was less than broadcast on the ATIS.

Trying to understand the language of aviation and all its acronyms is difficult for the nonpilot when there are no distractions; sometimes the task is impossible for a nonpilot when hearing the words over a radio from a fast-talking controller or other pilot. Even understanding the words doesn't necessarily mean understanding the meaning.

Take the sentence that starts this glossary, for instance. Confusing? With the explanations in this glossary you will decipher the code.

The following are just a few of the hundreds of terms used in aviation. Most of these definitions are culled from the *Pilot/Controller Glossary* published in the Airman's Information Manual by the Federal Aviation Administration, with an occasional line of explanation added by me. (As the FAA's glossary title implies, the definitions are also used by the air traffic controllers, which ensures understanding on both sides of the microphones.)

abeam An aircraft is "abeam" a fix, point, or object, when that fix, point, or object is approximately 90 degrees to the right or left of the aircraft track. Abeam indicates a general position rather than a precise point.

acknowledge Let me know that you have received my message.

advisory service Advice and information provided by a facility to assist pilots in the safe conduct of flight and aircraft movements.

aeronautical beacon A visual navigational aid displaying flashes of white and/or colored light to indicate the location of an airport, a heliport, a landmark, a certain point of a prescribed airway in mountainous terrain, or an obstruction.

air defense identification zone The area of airspace above land or water, extending upward from the surface, within which the ready identification, the location, and the control of aircraft are required in the interest of national security.

airmet In-flight weather advisories issued only to amend the area forecast concerning weather phenomena that are of operational interest to all aircraft and potentially hazardous to aircraft having limited capability because of lack of equipment, instrumentation, or pilot qualifications. Airmets cover moderate icing (that accumulates on an airplane), moderate turbulence, sustained winds of 30 knots or more at the surface, widespread areas of cloud ceilings that are less than 1,000 feet off the ground and/or visibility less than 3 miles, and extensive mountain obscurement.

airport advisory area The area within 10 miles of an airport without a control tower or where the tower is not in operation, and on which a flight service station is located.

airport elevation The highest point of an airport's usable runways measured in feet from mean sea level.

approach control service Air traffic control service provided by an approach control facility for arriving and departing VFR/IFR aircraft and, on occasion, en route aircraft. At some airports not served by an approach control facility, the air route traffic control center provides limited approach control service.

approach speed The recommended speed used by pilots when making an approach to landing. This speed, which is found in aircraft manuals, will vary for different segments of an approach, as well as vary for aircraft weight and the position of flaps and retractable landing gear.

ATC advises A phrase that is used to prefix a message of noncontrol information when it is relayed to an aircraft by other than an air traffic controller.

automatic altitude reporting That function of a transponder that responds to specific interrogations (Mode C) by transmitting the aircraft's altitude in 100-foot increments.

automatic direction finder An aircraft radio navigation system that senses and indicates the direction of a nondirectional radio beacon (NDB) ground transmitter. Direction is indicated to the pilot as a magnetic bearing or as a relative bearing to the longitudinal axis of the aircraft depending on the type of indicator installed in the aircraft. The direction finder can be tuned to receive regular AM stations when the receiver is not required for navigation.

automatic terminal information service The continuous broadcast of recorded noncontrol information in selected airport terminal area. Its purpose is to improve controller effectiveness and to relieve frequency congestion by automating the repetitive transmissions of essential but routine information.

cleared as filed Means the aircraft is cleared to proceed in accordance with the route of flight filed in the flight plan.

cleared for takeoff An air traffic controller's authorization to a pilot for an aircraft to depart. It is predicated on known traffic and known physical airport conditions.

cleared to land An air traffic controller's authorization to a pilot for an aircraft to land. It is predicated on known traffic and known physical airport conditions.

compass rose A circle, graduated in degrees, printed on some charts or marked on the ground at an airport. It is used as a reference to either true or magnetic direction.

DF guidance Headings provided to aircraft by facilities equipped with direction finding equipment. These headings, if followed, will lead the aircraft to a predetermined point such as the DF station or an airport. DF guidance is given to aircraft in distress or to other aircraft that request the service.

distance measuring equipment Airborne and ground electronic equipment that is used to measure, in nautical miles, the distance of an aircraft from the DME navigational aid.

dead reckoning Dead reckoning, as applied to flying, is the navigation of an airplane solely by means of computations based on airspeed, course, heading, wind direction and speed, groundspeed, and elapsed time.

decision height With respect to the operation of aircraft, means the height at which a pilot must make a decision during a precision instrument approach to either continue the approach or to execute a missed approach.

en route flight advisory service A service specifically designed to provide, upon pilot request, timely weather information pertinent to the type of flight, intended route of flight, and altitude. Pilots refer to the service as "flight watch" when calling on the radio.

fly heading (degrees) Informs the pilot of the heading that should be flown. The pilot might have to turn toward, or continue on, a specific compass direction in order to comply with the instructions. The pilot is expected to turn the shorter direction to the heading unless otherwise instructed by ATC.

handoff An action taken to transfer the radar identification of an aircraft from one controller to another if the aircraft will enter the receiving controller's airspace and radio communications with the aircraft will be transferred.

have numbers Used by pilots to inform ATC that they have received runway, wind, and altimeter information only.

instrument flight rules Rules governing the procedures for conducting instrument flight, typically when visibility is poor. Also a term used by pilots and controllers to indicate type of flight plan.

instrument landing system (ILS) A precision instrument approach system that normally consists of the following electronic components and visual aids: localizer, glideslope, outer marker, middle marker, and approach lights.

negative "No," or "permission not granted," or "that is not correct."

negative contact Used by pilots to inform ATC that: (1.) Previously issued traffic (a report from the controller that an airplane is nearby) is not in sight. The comment might be followed by the pilot's request for the controller to provide assistance in avoiding the traffic. (2.) They were unable to contact ATC on a particular frequency.

nonprecision approach A standard instrument approach procedure in which no electronic glideslope is provided.

notice to airmen A notice containing information (not known sufficiently in advance to publicize by other means) concerning the establishment, condition, or change in any component (facility, service, procedure) or hazard in the national flight operations.

phonetic alphabet The use of words to denote letters for clarity of communication. The following is the international phonetic alphabet of the International Civil Aviation Organization. English is the official language of aviation worldwide.

A	Alfa	**M**	Mike	**Y**	Yankee
B	Bravo	**N**	November	**Z**	Zulu
C	Charlie	**O**	Oscar	**0**	Zero
D	Delta	**P**	Papa	**1**	Wun
E	Echo	**Q**	Quebec	**2**	Too
F	Foxtrot	**R**	Romeo	**3**	Tree
G	Golf	**S**	Sierra	**4**	Fow-er
H	Hotel	**T**	Tango	**5**	Fife
I	India	**U**	Uniform	**6**	Six
J	Juliet	**V**	Victor	**7**	Sev-en
K	Kilo	**W**	Whisky	**8**	Ait
L	Lima	**X**	X-ray	**9**	Niner

prohibited area An airspace of defined dimensions, above the land areas or territorial waters of a state, within which flight of aircraft is prohibited.

runway visual range The distance along a runway that a pilot can see when fog or other conditions reduce visibility.

radar A radio detection device that provides information on range, azimuth, and/or elevation of objects. Additionally, *primary radar* is a radar system that uses reflected radio signals. *Secondary radar* is a radar system wherein a radio signal that is transmitted from a radar station initiates the transmission of a radio signal from another station. (The "other station" usually is the transponder in the aircraft. This provides a better radar return to the controller and also transmits additional information such as a discrete identification code and, in many cases, altitude information.)

radar approach An instrument approach procedure that utilizes precision approach radar (PAR) or airport surveillance radar (ASR).

radar contact Used by ATC to inform an aircraft that it is identified on the radar display and that radar flight following will be provided until radar identification is terminated. Radar service might also be provided within the limits of necessity and capability.

radar contact lost Used by ATC to inform a pilot that radar data used to determine the aircraft's position is no longer being received, or is no longer reliable and radar service is no longer being provided. The loss might be attributed to several factors including the aircraft merging with weather or ground clutter, the aircraft operating below radar line-of-sight coverage, the aircraft entering an area of poor radar return, failure of the aircraft transponder, or failure of the ground radar equipment.

runway heading The magnetic direction that corresponds with the runway centerline extended, not the painted runway number. When cleared to "fly runway heading," pilots are expected to fly or maintain the heading that corresponds with the extended centerline of the departure runway. When departing from Runway 4, which has a centerline magnetic heading of 044 degrees, fly 044.

say again Used to request a repeat of the last transmission. Usually specifies transmission or portion thereof not understood or received: "Say again all after Abram VOR."

say heading Used by ATC to request an aircraft heading. The pilot should state the actual heading of the aircraft.

see and avoid A visual procedure wherein pilots of aircraft flying in visual meteorological conditions (VMC), regardless of type of flight plan, are charged with the responsibility to observe the presence of other aircraft and to maneuver their aircraft as required to avoid the other aircraft.

special VFR conditions Meteorological conditions that are less than those required for basic VFR flight in Class B, C, D, or E surface areas and in which some aircraft are permitted flight under visual flight rules.

squawk (mode, code, function) A request from an air traffic controller for the pilot to activate a specific mode, code, and function on the aircraft transponder: "Squawk three/alpha, two one zero five, low."

standard instrument departure A preplanned instrument-flight-rule (IFR) air-traffic-control departure procedure printed for pilot use in graphic and/or textual form. SIDs provide a transition from the airport's terminal area to the appropriate en route portion of flight.

stand by Means the controller or pilot must pause for a few seconds, usually to attend to other duties of a higher priority. Also means to wait, as in "stand by for clearance." The caller should reestablish contact if a delay is lengthy. "Stand by" is not an approval or denial.

taxi into position and hold Used by an airport tower controller to inform a pilot to taxi onto the departure runway, stop in a takeoff position, and hold (wait). It is not authorization for takeoff. It is used when takeoff clearance cannot be immediately issued because of traffic that would obstruct a safe takeoff path or other reasons.

transponder A radio transmitter onboard an aircraft used to enhance and code an aircraft's radar return. Ground radar "interrogates" the airborne unit, and the signal return helps controllers to identify and tag the aircraft on video display radar screens. "Mode C" transponders also transmit the aircraft's altitude.

transcribed weather broadcast A continuous recording of meteorological and aeronautical information that is broadcast on aeronautical radio facilities for pilots.

uncontrolled airport An airport without an air traffic control tower. Of about 13,000 airports in the United States, fewer than 700 have towers. Uncontrolled does not mean hazardous flight. Pilots must fly prescribed patterns entering and leaving the airport and, at most locations, announce their positions and intentions over a unicom frequency.

visual flight rules Rules and regulations of the Federal Aviation Administration that govern flight by visual reference to the horizon and other outside indicators. Aircraft may fly VFR only when meteorological conditions are better than certain minimums. Pure jet aircraft are not permitted to operate VFR, but all others may, resulting in about 80 percent of all flights going under these rules.

wake turbulence Phenomena resulting from the passage of an aircraft through the atmosphere. The term includes disturbances of the air that can be quite violent in the air or on the ground: vortices, thrust-stream turbulence, jet blast, jet wash, propeller wash, and rotor wash. Pilots of small planes are especially wary of wake turbulence that is generated by an airliner during takeoff and landing.

waypoint A predetermined geographical position (oftentimes precise latitude and longitude coordinates) that is defined by reception of signals from electronic navigational aids: VORs, satellites, or other transmitters.

when able When used in conjunction with ATC instructions, this gives the pilot the opportunity to delay compliance until a condition or event has been reconciled; however, the pilot is expected to seek the first opportunity to comply. When a maneuver has been initiated, the pilot is expected to continue until the specifications of the instructions have been met.

These are but a fraction of the full glossary of terms used in aviation. Perhaps these definitions have introduced terminology that will prompt you to ask questions or do further research in the *Pilot/Controller Glossary* for a definition.

The right-seat companion is not expected to know or understand all aviation terminology, but a few terms, such as those in this glossary that are the most frequently used, will help take some of the mystery out of radio conversation. The terms and definitions also will give a clue to the conversations of pilots, helping to remove the glazed look of nonpilots when in the company of "hangar flying" pilots.

If you have decoded the sentence that started this glossary, try it out the next time a group of pilots leaves you wondering what the conversation is about. It will drive them crazy!

Index

Illustration page numbers are in **boldface.**

About the author

Charles Spence started in aviation in 1954 flying Piper Cub aircraft from a Long Island airport. He has been involved in aviation ever since, not as a professional pilot but using communication skills.

Spence began his career as a cartoonist in the promotion department of newspapers in Louisville, Kentucky. He worked on newspapers in Cincinnati and held executive positions on publications in Oakland, Calif., San Francisco, and New York City. He became a public relations executive in New York City and later made the flying avocation a vocation.

He was senior vice president of public relations for the Aircraft Owners and Pilots Association for 15 years. Spence reports from Washington, D.C., for several aviation publications and writes freelance articles that appear in aviation and general audience publications. He has received several national awards from the Aviation/Space Writers Association.

Spence maintains a current private pilot's license with single, multiengine, and instrument ratings.

Other Bestsellers of Related Interest

The Joy of Flying, 3rd Edition
Rob Mark
A quick and easy introduction for beginners to the requirements, challenges, and rewards of becoming a pilot. Revised to include updated Practical test Standards, new airspace designations, and expanded coverage of ratings and radio equipment.
0-07-040487-9 $15.95 Paper

Stick and Rudder
Wolfgang Langewiesche
Continuously in print for over 50 years it remains to be the leading thin-book on the art of flying. Applicable to large airplanes and small, old airplanes and new, it is of interest to the learner as well as the accomplished pilot.
0-07-036240-8 $19.95 Hard

Your Pilot's License, 5th Edition
Joe Christy, Revised and Updated by Jerry A. Eichenberger
The definitive guide to the daunting process of getting a pilot's license, now updated and expanded to include changes in regulations, testing, procedures, new airspace designations, and more.
0-07-019281-2 $14.95 Paper

Fly for Less: Flying Clubs and Aircraft Partnerships
Geza Szurovy
This valuable guide shows budget-minded pilots how to fly more and spend less through shared aircraft ownership. It's a pilot's legal and financial handbook, covering every aspect of shared ownership arrangements in clubs or among two or more pilots: financing, scheduling, insurance, maintenance, storage, liability, and more. Sample forms and Excel spreadsheet programs help simplify recordkeeping.
0-07-062858-0 $24.95 Hard

9 780070 601482